M12/A MA

Modeling the Impact of Climate Change on Rice Production in Asia

Edited by

Robin B. Matthews,
Martin J. Kropff,
Dominique Bachelet,
and

H.H. van Laar

CAB INTERNATIONAL
in association with the
International Rice Research Institute

CAB INTERNATIONAL
Wallingford
Oxon OX10 8DE
UK

Tel: +44 (0)1491 832111
Telex: 847964 (COMAGG G)
E-mail: cabi@cabi.org
Fax: +44 (0)1491 833508

Published in association with:

International Rice Research Institute
P.O. Box 933
1099 Manila
The Philippines

A catalogue entry for this book is available from the British Library.

ISBN 0 85198 959 4

Disclaimer: The information in this document has been funded
wholly or in part by the U.S. Environmental Protection Agency
under cooperative agreement number 817426 to the
International Rice Research Institute and contract number
68-C8-0006 to ManTech Environmental Technology Inc. It has
been subject to the agency's peer and administrative review. It has
been approved for publication as an EPA document.

Typeset by MFK Information Services Ltd, Hitchin, Herts.
Printed and bound in the UK by Biddles Ltd, Guildford.

Contents

Preface

The impact of global change on agriculture has been studied extensively for various crops at many different scales. However, relatively few studies have focused on rice, the most important staple food crop in Asia, where the already large population of rice consumers is rapidly increasing. In 1989, the United States Environmental Protection Agency's (EPA) Environmental Research Lab in Corvallis (ERL-C), Oregon, and the International Rice Research Institute (IRRI) in the Philippines initiated a major research project to investigate relationships between climate change and rice production. One component of this project was the quantification of the impact of climate change on rice production using crop simulation models, integrating existing knowledge of effects of increased levels of CO_2 and temperature.

This simulation component, hereafter referred to as Simulation of the Impact of Climate Change (SIC2), was partially funded by EPA, but executed by IRRI in collaboration with four teams of the SARP project (Simulation and Systems Analysis for Rice Production, a collaborative project of IRRI, the Research Institute for Agrobiology and Soil Fertility (AB-DLO), Wageningen, The Netherlands, and the Department of Theoretical Production Ecology at Wageningen University (TPE-LUW), The Netherlands), the Crop Science Laboratory at Kyoto University, Japan, and ERL-C.

The project was designed under the leadership of Dr F.W.T. Penning de Vries (IRRI, current address: AB-DLO, Wageningen, The Netherlands) and Dr Moon Hee Lee (Crop Experiment Station, Suweon, South Korea) during a workshop held in November 1990 at the China National Rice Research Institute (CNRRI), immediately following a SARP workshop on simulation of potential production of rice. Seven SARP teams who had expressed interest in participating in SIC2 were invited to the workshop, of which four were selected to take part in the project. The Crop Science Laboratory of Kyoto University in Japan, and ERL-C, also participated. During this workshop and a subsequent one held in 1992 in Japan, workplans were developed and tasks distributed: modifying and evaluating crop models, collection of historic weather data, simulation of future weather, simulation of impact of the climate change scenarios by country and summarizing results over Asia using Geographical Information Systems (GIS). The work started in January 1992 and finished at the end of 1993, resulting in this book.

R.B. Matthews
M.J. Kropff
D. Bachelet
H.H. van Laar

Los Baños/Corvallis
1 July 1994

Executive Summary

- The likely effects of climate change on rice production in east and southeast Asia were evaluated using a simulation and systems analysis approach at three different levels:

  1. Geographical Information System (GIS) data was used to estimate the effect of climate change on changes in the area and distribution of rice-growing areas.

  2. Two process-based crop simulation models, ORYZA1 and SIMRIW, calibrated for *indica* and *japonica* ecotypes, were used to predict changes in rice production of the whole region. Both models simulate the development and potential growth of rice in relation to temperature, solar radiation, CO_2, and characteristics of the varieties used.

  3. The same two crop models, calibrated for local varieties and conditions in a number of Asian countries, were used to evaluate likely changes in a more detailed analysis within each country.

- In the GIS analysis, sixteen agroclimatic zones in the region were defined according to current annual temperatures and rainfall. Climate change scenarios predicted by the General Fluid Dynamics Laboratory (GFDL), the Goddard Institute of Space Studies (GISS), and the United Kingdom Meteorological Office (UKMO) General Circulation Models (GCMs) were then superimposed. All scenarios predicted a rise in temperatures in all zones, and an increase in precipitation in 60% of the area. Significant geographic redefinition of some of the zones was predicted, but it was suggested that this would not necessarily be matched by a corresponding shift in rice-growing areas due to physical geographic barriers.

- In the crop simulation analysis, two approaches were used. Firstly, fixed increments of CO_2 level and temperature above current levels were used in various combinations to investigate the sensitivity of the Asian rice production system to changes in these two variables. Secondly, climate change scenarios predicted by the GFDL, GISS, and UKMO GCMs were used to analyze likely changes in the potential yield and production of rice in future climates.

- It was assumed that proportional changes in potential yields reflected the proportional changes in actual yields. Both crop models predicted decreases in yield with an increase in temperature alone, and increases in yield when only CO_2 level was raised. However, the exact effect on existing crop yields at a specific site depended on the combination of these two factors at that site. The uncertainty with which these can be predicted in future climate scenarios places a limit on the accuracy of estimation of the likely effect of these changes on crop yields.

- Using the ORYZA1 model, overall rice production in the region was predicted to change by +6.5%, −4.4%, and −5.6% under the GFDL, GISS, and UKMO doubled-CO_2 scenarios respectively. The corresponding changes predicted by the SIMRIW model were +4.2%, −10.4%, and −12.8%. The average of these estimates would suggest that rice production in the Asian region may decline by −3.8% under the climates of the next century. For both crop models, the predicted changes varied considerably between countries, with declines in yield predicted under the GISS and UKMO scenarios for Thailand, Bangladesh, and western India, while increases were predicted for Indonesia, Malaysia, and Taiwan, and parts of India and China. Such changes would probably have a significant effect on future trading relationships within the region.

- In Japan, it was predicted that, although there would be no change under a doubled-CO_2 climate in overall rice production, there would be a significant change in the distribution of this production. Northern areas would benefit from longer growing seasons and less yield variability caused by cool summer damage, while southern areas would suffer yield declines and increased yield variability from more frequent occurrences of spikelet sterility due to high temperature.

- In India, annual national rice production was predicted to increase under the three GCM scenarios, mainly due to an increase in yields of main season crops where the effect of higher CO_2 levels was more than able to compensate for the detrimental effects of higher temperatures. Although large yield decreases were predicted for second season crops at many of the locations due to high temperatures being encountered, the small fraction of total rice produced in this season meant that its effect on overall national production was small.

- In Malaysia, increases in annual production of 20–30% were predicted for the three scenarios. This was because temperatures during the growing seasons in the current climate are about 26°C, which, even with the temperature increments predicted by the GCMs, do not rise to a level where spikelet fertility is likely to be influenced by high temperatures. The increased temperatures also did not shorten the duration of the crop sufficiently to negate the beneficial effect of the increased CO_2 level.

- In South Korea, changes of −9%, 1%, and −25% in annual national production were predicted for the GFDL, GISS, and UKMO scenarios respectively, suggesting that, overall, a decline in production may occur. Detailed analysis showed that this decline was mainly due to a decrease in yields in the lower altitude areas, where temperatures at the time of flowering were high, resulting in spikelet sterility. The effect of increased temperature was less marked at higher altitudes.

- In China, annual rice production was predicted to change by +12.2%, +2.0% and +5.6% under the GFDL, UKMO, and GISS scenarios respectively, although those changes varied considerably between seasons. 'Early' and 'late' rice were predicted to have yield increases, as high temperatures around the time of flowering were avoided, but 'single' rice in many of the locations was predicted to decline, primarily due to significant spikelet sterility. In many cases, yield declines were also accompanied by an increase in the variability of yields.

- In the Philippines, national rice production was predicted to change by +6.6%, −14.0%, and +1.1% for the GFDL, GISS, and UKMO scenarios. These changes were mainly due to the effect on wet season production, which contributes 80% towards current annual production. Changes predicted in the dry and transition seasons, while large in some cases, were not so significant due to their lesser contribution to overall production. Nevertheless, a warmer climate may allow a third rice crop to be grown in higher altitude regions where only two crops can be grown at present.

- The response of spikelet sterility to temperature emerged as a major factor determining the differential predictions of the effects of climate change on each scenario. Due to the extreme sensitivity of this factor to temperatures below 20°C and above 33°C, a small change in mean temperature in these ranges can result in a large change in yield, due to the effect on the numbers of grains formed.

- Varietal adaptation, such as tolerance to high temperatures, was shown to be capable of ameliorating the detrimental effect of a temperature increase in currently high-temperature environments. For spikelet fertility, the level of adaptation required is within the range of genotypic variation currently observed.

- Changes in management practices, such as advanced planting dates in the higher latitudes, could give increased yields, but the use of longer-maturing varieties at such sites, as suggested by some authors, may result in lower yields, due to the grain formation and ripening periods being pushed to less favorable conditions later in the season.

List of Abbreviations

AB-DLO	Research Institute for Agrobiology and Soil Fertility, Wageningen, The Netherlands
AET	Actual evapotranspiration
AEZ	Agroecological zone
BVP	Basic vegetative period
CES	Crop Experiment Station, Suweon, South Korea
CGR	Crop growth rate
CLICOM	CLImate COMputing system developed by the World Meteorological Organization
CNRRI	China National Rice Research Institute, Hangzhou, China
CSD	Cool-summer damage to spikelets
DAS	Days after sowing
Dd	Developmental day
DRLVT	Relative death rate of the leaves as a function of development stage
DSSAT	IBSNAT Decision Support System for Agrotechnology Transfer
DVI	Development index (SIMRIW model)
DVII	Initial value of development index (SIMRIW model)
DVR	Developmental rate (ORYZA1 model)
DVS	Development stage (ORYZA1 model)
EPA	United States Environmental Protection Agency
ERL-C	EPA Environmental Research Lab in Corvallis
FAO	Food and Agriculture Organization of the United Nations
FLVTB	Fraction of dry matter partitioned to the leaves
FSHTB	Fraction of dry matter partitioned to the shoot
FSTTB	Fraction of dry matter partitioned to the stem
GCM	General Circulation Model
GDP	Gross Domestic Product
GFDL	Geophysical Fluid Dynamics Laboratory GCM model
GFP	Grain-filling phase

GHG	Greenhouse gases
GIS	Geographic Information System
GISS	Goddard Institute for Space Studies GCM model
HYV	High-yielding variety
IBSNAT	International Benchmark Sites Network for Agrotechnology Transfer
IDOYS	Date of sowing
IDOYTR	Date of transplanting (equals date of seeding for direct-seeded rice)
IPCC	Intergovernmental Panel on Climate Change
IRRI	International Rice Research Institute, Los Baños, The Philippines
LAI	Leaf area index
LAII	Initial leaf area index
LLV	Loss of the leaves
MACROS	Modules of an Annual Crop Simulator (Wageningen crop models)
NCAR	National Center for Atmospheric Research, USA
NFLVTB	Areal leaf nitrogen content as a function of development stage
NH	Number of hills per m^2
NPLH	Number of plants per hill
NPLSB	Approximate number of plants per m^2 in seedbed
PAI	Panicle area index
PAR	Photosynthetically active radiation
PET	Potential evapotranspiration
PFP	Panicle formation phase
PI	Panicle initiation
PPT	Annual precipitation
PSP	Photoperiod-sensitive phase
RDA	Rural Development Administration, South Korea
RDTT	Daily solar radiation as function of the day of year
RGRL	Relative growth rate of leaf area as a function of thermal time
RLDR	Relative leaf death rate as a function of development stage
RUE	Radiation use efficiency
SAI	Stem (or sheath) area index
SARP	Simulation and Systems Analysis for Rice Production Project
SIC2	Simulation of the Impact of Climate Change Project
SIMRIW	SImulation Model for RIce–Weather relations
SLA	Specific leaf area (area of leaf per unit leaf weight)
SLATB	Specific leaf area as a function of development stage
SOW	Sowing date (julian day)
SSGA	Specific green stem area (area of stem per unit stem weight)
SUMDD	Sum of degree-days
TMAXT	Maximum daily temperature (°C) as function of the day of year
TMINT	Minimum daily temperature (°C) as function of the day of year
TNRRI	Tamil Nadu Rice Research Institute, Adutherai, India
TPE-LUW	Department of Theoretical Production Ecology, Wageningen University, The Netherlands
TPLT	Transplanting age (days after sowing)
UKMO	United Kingdom Meteorological Office GCM model
UPM	Universiti Pertanian Malaysia, Serdang, Malaysia
WMO	World Meteorological Organization

CLIMATE CHANGE AND RICE
I

Introduction 1

R.B. Matthews[1], M.J. Kropff[1], and D. Bachelet[2]

[1]*International Rice Research Institute, P.O. Box 933, 1099 Manila, The Philippines;* [2]*ManTech Environmental Technology, Inc., US EPA Environmental Research Laboratory, 200 SW 35th Street, Corvallis, Oregon 97333, USA*

Although rapid industrialization over the last century has contributed to a major improvement in the living standards of millions of people, it has also brought with it serious problems. Industrial and agricultural emissions of carbon dioxide (CO_2), methane (CH_4), chlorofluorocarbons, nitrous oxide, and other gases, mainly from the burning of fossil fuels, are resulting in an increase of these gases in the earth's atmosphere; CO_2 concentration, for example, is currently increasing at the rate of about 1.5 ppm year^{-1} (Keeling *et al.*, 1984). This increase in these so-called 'greenhouse gases' (GHGs) is contributing to a gradual warming of the planet by retaining more heat within the earth's atmosphere (Gates *et al.*, 1992). General Circulation Models (GCMs), describing the dynamic processes in the earth's atmosphere, have been used extensively to provide potential climate change scenarios (e.g. Grotch, 1988; Gutowski *et al.* 1988; Smith & Tirpak, 1989; Cohen, 1990). Based on current rates of increase it is generally accepted that CO_2 levels will reach double the current level by the end of the next century. According to current GCM predictions, a doubling of the current CO_2 level will bring about an increase in average global surface air temperatures of between about 1.5 and 4°C, with accompanying changes also in precipitation patterns.

The relationship between climatic changes and agriculture is a particularly important issue, as the world's food production resources are already under pressure from a rapidly increasing population. Both land use patterns and the productivity of crops are likely to be affected (Solomon & Leemans, 1990); it is vital, therefore, to obtain a good understanding, not only of the processes involved in producing changes in the climate, but also the effect of these changes on the growth and development of crops.

Rice is the second most important crop in the world after wheat, with about

522 million tonnes being produced from about 148 million hectares in 1990. It is grown over a considerable geographic range from 45°N to 40°S to elevations of more than 2500 m, but with average daily temperatures in the range of 20–30°C (Oldeman *et al.*, 1987). Cold temperatures currently limit rice production in cooler areas of the temperate regions and mountainous areas. The largest production of rice is from Asia, which produces about 94% of the total world production. In this region, which also contains 59% of the global population, rice is the main item of the diet, and provides an average of 35% of the total caloric intake compared to only 2% in the USA. Table 1.1 shows the current production, areas planted, mean yields, and quantities of rice exported and imported within the countries in this study. However, to meet the demands of its rapidly expanding population, an estimated 70% increase in rice production is required over the coming decades (IRRI, 1993). This is a major challenge for

Table 1.1. Production of rough rice, area planted, and country-level yields in 1990, and imports and exports in 1990 for the countries used in this study. (Source: IRRI, 1991.)

Country	Production ('000 t)	Area ('000 ha)	Yield (t ha^{-1})	Imports ('000 t)	Exports ('000 t)
Bangladesh	29,400	10,600	2.77	53	–
China	188,300	32,900	5.72	1,206	384
India	109,500	41,800	2.62	564	300
Indonesia	43,846	10,000	4.38	397	–
Japan	12,500	2,095	5.97	16	–
Malaysia	1,800	628	2.87	367	–
Myanmar	13,623	4,661	2.92	–	159
Philippines	9,600	3,525	2.72	195	16
South Korea	7,600	1,250	6.08	2	1
Taiwan	2,300	500	4.60	3	68
Thailand	20,400	10,200	2.00	–	6,311

the agricultural research community and policy-makers alike, particularly as yields in some of the more productive farmers' fields are already approaching the ceiling of average yields obtained on experimental stations (IRRI, 1993), thereby minimizing gains in production through improved management practices. The likely impact of climate change on rice production, therefore, only adds to an already complex problem, and is of paramount importance in planning strategies to meet the increased demands for rice in the next century.

In recent years, a number of controlled-environment studies have added to our understanding of the effect of increased temperature and CO_2 on the production of many crops (see Kimball, 1983; Acock & Allen, 1985, for reviews). In rice, a doubling of the CO_2 level from 340 ppm to 680 ppm is predicted to increase productivity and yields on average by 30% under optimum conditions

(Kimball, 1983), mainly through the stimulation of photosynthetic processes in the plant (Lemon, 1983; Cure & Acock, 1986) and improvement in the water use efficiency (Goudriaan & van Laar, 1978; Gifford, 1979; Sionit *et al.*, 1980). The effect of temperature is more complex, however. Crop duration decreases as temperatures rise to 30°C, then increases as temperature rises still further; this affects the length of time the crop has to produce grains and fill them. Below 20°C and above about 32°C, spikelet sterility becomes a major factor, even if there is sufficient growth in other plant components. Interactions between these effects are also important; a temperature increase at higher latitudes, for example, while generally decreasing yields by reducing crop duration, may also decrease spikelet sterility from cold damage, more than offsetting any detrimental effect. The overall effect of an increase in CO_2 and temperature is complicated, therefore, and depends on the relative effects of both variables in the particular combination occurring for a given region. Estimating the effect of a changing climate on rice production in the Asian region is made even more difficult due to the variety of cropping systems and levels of technology used. However, the use of crop growth models is one way in which these effects can be studied, and probably represent the best method we have at present of doing so. Although a large number of simplifying assumptions must necessarily be made, these models allow the complex interaction between the main environmental variables influencing crop yields to be understood.

A number of crop modeling studies have emerged in recent years of the likely effects of climate change on rice production. Several of these so far have been limited to single countries or subregions within countries (e.g. Bangladesh: Karim *et al.*, 1991; Japan: Horie, 1988; China: Zhou, 1991), and many have only considered the effect of temperature changes without including the influence of CO_2 (e.g. Okada, 1991). Others are based on statistical regression models only (e.g. Wang *et al.*, 1991). Yoshino *et al.* (1988a) predicted that lowland rice yields could increase in Japan by about 9% following a doubling of CO_2 and subsequent climatic changes as predicted by the Goddard Institute of Space Studies (GISS) GCM (Hansen *et al.*, 1988). Solomon & Leemans (1990), using a very simple model and long-term monthly-average climatic data in a world-wide study, predicted a yield increase of 0.4% for the current rice-growing environments, but little change in the areas sown because of the sharp temperature and moisture gradient along the northern border of its primary distribution in eastern Asia.

Jansen (1990) used historic weather data from seven sites in eastern Asia and the MACROS crop simulation model (Penning de Vries *et al.*, 1989) to evaluate the possible effect of various climate change scenarios on regional rice production. Simulated yields rose when temperature increases were small, but declined when temperatures increased more than 0.8°C per decade, with the greatest decline in crop yields occurring between the latitudes of 10 and 35°. Similar results were obtained by Penning de Vries (1993) and Kropff *et al.* (1993). The effects were the result of increased photosynthesis at higher CO_2

and a reduced length of the growing season at higher temperatures. Scenarios predicted by GCMs, however, were not considered. Rosenzweig *et al.* (1993), working with collaborators from 22 countries, used a number of the IBSNAT crop models to simulate likely changes in production of various crops under various GCM scenarios. They predicted that crop yields are likely to decline in the low-latitude regions, but could increase in the mid and high latitudes. The different responses were related to the current growing conditions; at low latitudes, crops are currently grown nearer their limits of temperature tolerance, so that any warming subjects them to higher stress, whereas in many mid- and high-latitude areas, increased warming benefited crops currently limited by cold temperatures and short growing seasons. Again, only a limited number of sites (21 for rice in east and south-east Asia) were used. Leemans & Solomon (1993) used a simple model based on temperature, solar radiation and rainfall data in a Geographical Information System (GIS) environment to estimate the effects on the production of various crops on a global scale and predicted an 11% increase in global production of rice. Their model did not, however, include physiological responses specific to rice, such as the response of spikelet fertility to high and low temperatures.

From these various estimates, it is clear that a wide range of predictions have been made on the likely effect of climate change on the production of rice. These studies contain many uncertainties, partly due to uncertainties in the predictions of the GCMs themselves, partly from the use of limited sites for which historical weather data is available, and partly from the quality of the crop simulation models, especially for rainfed conditions (Bachelet *et al.*, 1993). Although monthly average weather data are more easily available, the availability of good historical daily weather data is crucial to predict yield variability using ecophysiological simulation models, as many of the physiological relationships used are non-linear (Nonhebel, 1993).

The use of simulation models to predict the likely effects of climate change on crop production is, of necessity, an evolving science. As both general circulation models and crop simulation models become more sophisticated, and as more high-quality historical weather data for a larger number of sites becomes available, predictions will become more accurate. Revised predictions have already occurred; Horie (1991), for example, calculated that there would be an overall decline in rice production in Japan under predicted climate change scenarios, but this study (Chapter 8) concludes that, while there will be a shift in the rice-producing regions within the country, the overall rice production of Japan will not be appreciably altered.

The present study, therefore, can be seen as part of this evolutionary process, and complimentary to previous studies. We have approached the problem at three different levels. Firstly, we have used GIS data to estimate the effect of climate change on changes in the distribution and size of rice-growing areas (Chapter 6). Secondly, we have used two different crop simulation models, both calibrated for the two main rice ecotypes, *indica* and *japonica*, to evaluate likely changes in rice production at the regional level (Chapter 7). Thirdly, we have used the same two crop models, calibrated for local varieties in a number of Asian countries, to evaluate likely changes in a more detailed analysis (Chapters 8–13).

The weather database that has been developed, containing historical daily weather data collected from some 68 sites in the rice-growing regions of 11 countries, is described in Chapter 2. The crop simulation models used, which incorporate the effect of supra-optimal temperatures on phenological development and the effect of high and low temperatures on spikelet fertility, are described in Chapters 3 and 4, while a description of the GCM scenarios and their limitations for impact studies are presented in Chapter 5. The study, therefore, represents the most detailed so far of the effects of climate change on rice production in the east and south-east Asian region.

The Rice–Weather Database 2

H.G.S. Centeno[1], A.P. Aclan[1], A.D. Balbarez[1], Z.P. Pascual[1], F.W.T. Penning de Vries[2], and M.J. Kropff[1]

[1] *International Rice Research Institute, P.O. Box 933, 1099 Manila, The Philippines;* [2] *Research Institute for Agrobiology and Soil Fertility, Bornsesteeg 65, 6700 AA Wageningen, The Netherlands*

2.1 Introduction

The potential production of a crop is assumed to be determined solely by the interaction of genotypic characteristics with the solar radiation, temperature, CO_2 level, and daylength that it experiences. Solar radiation provides the energy for the uptake of CO_2 in the photosynthetic process, while temperature determines the crop growth duration, and the rates of physiological and morphological processes. Daylength can affect the rate of development at certain phases of the crop's life cycle, and to a lesser extent, the amount of solar radiation received by the crop. Crop models simulating potential production of rice, therefore, need solar radiation, temperature, and CO_2 level as inputs. Daylength can be calculated from the latitude and day of the year. For the simulation of rainfed rice systems, additional information on rainfall, humidity, wind speed, and the hydraulic characteristics of the soil is needed.

In the current study, however, we have decided to evaluate the effects of climate change on overall rice production in terms of the proportional effects on potential production. It is recognized that this approach has limitations in that it assumes weeds, diseases, and insect pests are absent, water and fertilizers are in abundance, that there are no adverse soil conditions, and that no extreme weather events, such as typhoons, occur. Although this may not be completely realistic, the approach has been found to explain a large part of the current variation in farmers' yields from year to year due to weather (see Chapter 8). We have therefore assumed that the proportional changes in yield brought about by climate change are the same at the potential production level as at other levels of production, such as when water or nitrogen is limiting, or if pests or diseases are present. While this assumption is open to some debate, we feel that the approach is justified for two reasons. First, while it may be suggested that for a crop growing in suboptimal conditions (i.e. low nitrogen supply, water stress, or suffering pest or disease damage) the enhancing effect of higher levels of CO_2 would be proportionally less than under optimal conditions, there is very little data in the literature with which to quantify this assumption. Second, in relation to water-limited rice production, in view of the poor ability of General Circulation Models (GCMs) to predict even current precipitation patterns (Chapter 5; also Bachelet *et al.*, 1993), let alone patterns 100 years hence, we feel that our assumption of similar proportional changes in yield at all levels of production is likely to introduce less error than using water-limited rice simulation models directly. Moreover, as irrigated rice accounts for 76% of global rice production (IRRI, 1993), errors introduced by this assumption into overall estimates are not likely to be large.

In the following analysis, therefore, only values of minimum and maximum temperatures, solar radiation, and latitude are used as environmental inputs into the crop simulation models, ORYZA1 and SIMRIW, used.

2.2   Weather Data Collection

Daily weather data were collected from various rice-growing areas in Asia, either by the National Weather Bureaux in the different countries, by the Climate Unit at IRRI (ten stations in the Philippines), or by the participants of the IRRI-WMO Rice–Weather Project based at IRRI from 1984–1986, out of which five collaborating institutes continued to collect and supply weather data after the end of the project in 1986 (Oldeman *et al.*, 1986). The collaborators in the current IRRI-EPA project described in this report acquired copies of weather data from their respective countries from at least one station in the major agroecological zones where rice is grown. Wherever possible, ten or more years of historic weather data from 68 stations in 15 countries from rice-growing areas were compiled in the rice–weather database.

The meta-data of the database is given in Table 2.1. The basic station description includes station name, latitude, longitude, elevation, and year coverage. Latitude and longitude are specified to the nearest hundredth of a degree. The elevation is specified in meters as reported by the data collectors. Most stations provided only station names and coordinates, although for some, more detailed information, such as station relocations and changes in instrumentation, was available. The CLICOM system (CLImate COMputing system developed by the World Meteorological Organization (WMO), 1989) was used for data storage and data validation. The database is managed by the Climate Unit of IRRI.

Table 2.1.   Details of the weather stations used in the study showing the location, years of weather data available, longitude and latitude of the site, and the elevation.

Country	Station	Years	Longitude (+ °E)	Latitude (+ °N)	Elevation (m)
Bangladesh	Khulna	11 (1980–1990)	89.6	22.9	5
	Jessore	11 (1980–1990)	89.2	23.2	7
	Rangpur	11 (1980–1990)	89.3	25.8	32
	Joydebpur	10 (1983–1992)	90.4	23.9	8
China	Beijing	9 (1980–1988)	116.5	39.9	55
	Chansha	9 (1980–1988)	113.1	28.2	45
	Chendu	8 (1980–1986, 1988)	104.0	30.7	506
	Fuzhou	9 (1980–1988)	119.3	26.1	85
	Guangzhou	9 (1980–1988)	113.3	23.1	18
	Guiyang	9 (1980–1988)	106.7	26.6	1074
	Hangzhou	9 (1980–1988)	120.2	30.2	45
	Nanjing	9 (1980–1988)	118.8	32.1	9
	Wuhan	8 (1980–1986, 1988)	114.1	31.0	23
	Shenyang	9 (1980–1988)	123.4	41.8	42
India	Aduthurai	31 (1960–1965, 1968–1992)	79.5	11.0	19
	Coimbatore	6 (1983–1985, 1988–1990)	77.0	11.0	431
	Cuttack	6 (1983–1987, 1990)	86.0	20.5	23
	Hyderabad	2 (1983–1984)	78.4	17.4	545
	Kapurthala	3 (1983–1985)	75.9	30.9	247
	Patancheru	10 (1975–1984)	78.4	17.5	25
	Bijapur	10 (1972–1980)	75.8	16.8	594
	Madurai	6 (1986–1991)	78.0	8.5	147
	Pattambi	3 (1983–1985)	76.2	10.8	25
Indonesia	Bandung	8 (1980–1986, 1988)	107.5	−6.9	791
	Ciledug	8 (1981–1988)	106.8	−6.3	26
	Cimanggu	3 (1989–1992)	106.7	−6.6	240
	Cipanas	3 (1989–1991)	107.0	−6.8	1100
	Maros	15 (1975–1989)	119.5	−5.0	5
	Muara	8 (1981–1985, 1990–1992)	106.7	−6.8	260
	Pacet	4 (1989–1992)	107.0	−6.8	1138

Continued

Table 2.1. *Continued*

Country	Station	Years	Longitude (+ °E)	Latitude (+ °N)	Elevation (m)
Japan	Akita	12 (1979–1990)	140.1	39.7	9
	Hiroshima	12 (1979–1990)	132.4	34.4	29
	Kochi	12 (1979–1990)	133.5	33.6	2
	Miyazaki	12 (1979–1990)	131.4	31.9	7
	Maebashi	12 (1979–1990)	139.1	36.4	112
	Nagoya	12 (1979–1990)	137.0	35.2	51
	Sapporo	12 (1979–1990)	141.3	43.1	17
	Sendai	12 (1979–1990)	140.9	38.3	39
	Toyama	12 (1979–1990)	137.2	36.7	9
Malaysia	Kemubu	8 (1983–1990)	102.3	5.9	1
	Telok Chengai	11 (1978–1988)	100.3	6.1	1
	Tanjung Karang	11 (1980–1990)	101.2	3.5	2
Myanmar	Myananda	6 (1985–1990)	78.4	17.5	n/a
	Pyay	6 (1985–1990)	97.0	18.0	n/a
	Wakema	6 (1985–1990)	95.7	16.8	n/a
	Yezin	8 (1983–1990)	96.1	22.0	n/a
Philippines	Albay	17 (1972–1973, 1975–1989)	123.7	13.4	n/a
	Banaue	14 (1979–1992)	121.1	16.9	1040
	Betinan	12 (1980–1991)	123.5	7.9	45
	Butuan	10 (1981–1990)	125.5	8.9	46
	Cavinti	8 (1985–1992)	121.5	14.3	305
	Claveria	8 (1985–1992)	124.9	8.6	305
	CLSU	17 (1974–1990)	120.9	15.7	76
	CSAC	16 (1975–1990)	123.3	13.6	36
	Guimba	7 (1986–1992)	120.8	15.7	66
	H. Luisita	14 (1971–1974, 1978–1984, 1988–1990)	120.6	15.4	32
	Iloilo	30 (1961–1990)	122.6	10.7	305
	IRRI Dryland	14 (1979–1992)	121.3	14.2	21
	IRRI Wetland	14 (1979–1992)	121.3	14.2	21
	La Granja	16 (1975–1990)	122.9	10.4	84
	MMSU	13 (1976–1988)	120.6	18.0	5
	PNAC	13 (1977–1981, 1983–1990)	118.6	9.4	7
	UPLB	32 (1959–1990)	121.3	14.2	21

Continued

Table 2.1. *Continued*

Country	Station	Years	Longitude (+°E)	Latitude (+°N)	Elevation (m)
South Korea	Chongju	16 (1977–1990, 1991–1992)	127.4	36.6	59
	Ch'ilgok	12 (1977–1984, 1986–1989)	128.6	36.0	55
	Chinju	15 (1977–1989, 1991–1992)	128.1	35.2	22
	Ch'unch'on	14 (1977–1988, 1991–1992)	127.7	37.9	74
	Daejun	14 (1977–1988, 1991–1992)	127.4	36.3	77
	Kangnung	15 (1977–1989, 1991–1992)	128.9	37.8	26
	Kwanju	14 (1977–1988, 1991–1992)	126.9	35.1	71
	Chonju	13 (1977–1987, 1991–1992)	127.2	35.8	51
	Milyang	18 (1973–1988, 1991–1992)	128.7	35.5	12
	Suweon	15 (1977–1989, 1991–1992)	127.0	37.3	37
	Cheolweon	12 (1977–1988)	127.1	38.2	192
Taiwan	Pingtung	8 (1983–1990)	120.5	22.7	24
Thailand	Chiang Mai	4 (1975–1988)	99.0	18.8	313
	Khon Kaen	14 (1975–1988)	102.8	16.4	165
	Ubon Ratchathani	16 (1975–1990)	104.9	15.3	123

n/a = not available.

2.3 Data Quality

Data quality in the database depends on the quality, calibration, and maintenance of the instruments at the station, the accuracy of data reports, and the
procedures used for estimating missing data. Data evaluation and quality control have been performed at both the observation site and the local data center.
However, except for the stations that IRRI maintains in the Philippines, documentation about the instruments and their recalibration was not available.
Therefore, validation of historic weather data was generally conducted without direct feedback from the observation site.

The collected set of weather data was received in the form of either computer files, photocopies of observers' logbooks, or data reports. Copies of the original logbooks were collected whenever possible to verify the data. If only the observer's logbook was received, the data were tabulated into data reports first. The data reports were entered in CLICOM.

The data quality control process consisted of two phases: (i) initial validation using global standards; and (ii) detailed validation using information on station statistics. Upon receipt of the data, the available weather elements in the set were identified. Then the data were checked for:

- The number of observations per year
- The code for missing values
- The unit of measurement per element
- The number of digits in which each element is reported.

Whenever possible, information about the station and the instruments installed in the stations was collected. For example, some weather stations have anemometers installed at different heights. Wind velocity observations were corrected to 2 m height values using the following relationship (Baier & Russel, 1968, as cited by Asuncion, 1971):

$$u_2 = 0.301 \, u_1/\log z_1$$

where z_1 is the installation height of the anemometer, and u_1, u_2 are wind velocity at z_1 and 2 m height, respectively.

Data were input into CLICOM by direct key-entry, or through its data import facility. As data were keyed, each value was subjected to three quality control checks to avoid typing errors: value limits, rate of change limits, and consistency between elements. The three automatic checks were performed immediately after the data-entry operator entered a line of data. The criteria have been selected by the IRRI Climate Unit and indicate rough limits of values for the elements in different rice-growing environments. The criteria were based on the long-term values from Climatological Standard Normals published by the different National Meteorological Bureaux (e.g. Korea Meteorological Administration, 1991) or as compiled by the Food and Agriculture Organization of the United Nations (FAO, 1987). For example, the solar radiation limits are 0 and 40 MJ m^{-2} d^{-1}. If values were outside the indicated range, the operator was asked to reenter the value. If the same value was entered again, it was accepted. The rate of change criteria and the consistency criteria were preset as suggested by WMO. If a different value was entered, it was subject to the same quality control checks as the original. Therefore, any questionable value had to be keyed twice. This check avoided most typing errors.

All data stored into CLICOM were subject to the second validation process. In contrast to the key-entry validation, the second phase of quality control required a skilled validator to pass judgment on suspicious values. The following checks were conducted in this second validation phase. For the first five years of data the general criteria that were used in the key entry were used. For all additional data, station specific criteria were derived from the first five years. The following steps have been taken:

1. Checking if the value was within a reasonable range by comparing with the maximum and the minimum values ever recorded;
2. Checking against the previous value entered to ensure that it had not changed more than could be expected; and
3. Checking against other meteorological characteristics entered for the same day to make sure that they were not inconsistent.

The Climate Unit received about ten years of weather data per station. The third validation check was conducted by the validator when at least five years of weather data were loaded into CLICOM. From these, simple monthly means and extreme values were generated and used as bases to check the consistency of other incoming data. The WMO recommends 30 years of weather data as a basis for stable long-term values, although ten years would be a viable base to begin a climate database. As examples of data inconsistency, a large variation in annual means was found in Chunchon (South Korea) and in Los Baños (The Philippines) data set. In Chunchon, different units of measurement in global solar radiation were used in a single data file. Reported values in langleys are generally ten times higher than those in MJ m^{-2} and can easily be detected. In Los Baños, the annual mean global radiation decreased sharply from the 1960s until the 1980s. Upon looking at the observer's logbook, it seemed that the sharp decline was caused by an uncalibrated pyranometer since its installation in 1959, and a mistake in unit conversion of the daily data values from langleys to mWh cm^{-2} to MJ m^{-2}. This experience indicated the importance of proper documentation of weather instrumentation and data validation. Daily global radiation increases with the duration of bright sunshine and decreases with rainfall. Examination of the time series data for a summer month in South Korea revealed an isolated wet and cloudy day on 13 May. During that day, the reading from the radiometer was too low since the tolerance limit for a sound observation of that instrument was not met.

2.4   Estimating Missing Values

In CLICOM, questionable values are identified but not automatically changed. The system depends on the experience of the person who conducts the validation as to whether they accept or reject the values in question. In our case, this validation is performed by both a professional MSc agroclimatologist and statistician with agroclimatological experience. A questionable value can be accepted with a high or a low confidence, or can be replaced with an estimate. If a value is accepted with a low confidence, it is tagged as *D* for dubious. Estimated values are flagged as *E*, to differentiate them from actual observations.

Other flags include *G* for a generated or derived value, *T* for trace rainfall, and *M* for a missing value.

2.4.1 Solar radiation

Most solar radiation data were obtained by reading global solar radiation from a pyranometer. A sunshine recorder is a common backup instrument to measure the duration of bright sunshine. There are several ways to estimate missing values of solar radiation. In some locations, the duration of bright sunshine hours was recorded. Daily total global solar radiation (MJ m^{-2} day^{-1}) was estimated from the recorded duration of bright sunshine using the equation

$$RAD = RAD_a \{a + b \cdot (SSD/DAYL)\} \quad \text{(Angström, 1924)}$$

where RAD_a is the extraterrestrial radiation (MJ m^{-2} d^{-1}), SSD is the duration of bright sunshine (h), $DAYL$ is the daylength (h), and *a*, *b* are dimensionless coefficients that depend on the prevailing climatic conditions and latitude.

Indicative values for the empirical constants *a* and *b* in the Angström formula, in relation to latitude used by the FAO (Frère & Popov, 1979) are given in Table 2.2:

Table 2.2.   Values for constants *a* and *b* in the Angström formula.

Climate	*a*	*b*
Arid, semi-arid	0.25	0.45
Tropical humid	0.29	0.42
Others	0.18	0.55

For periods of up to three consecutive days with missing values, solar radiation data were estimated by interpolation. The missing value was estimated from the mean of observations from five days before and five days after the period without observations. When no data were available for periods with more than three consecutive days, or ten non-consecutive days of missing values in a month, readings were coded as −99. Interpolation of these data is then left for the user. In the crop simulation models used in the present study, the solar radiation data are estimated by linear interpolation for periods of up to ten days (see Chapters 3 and 4 for more details). Beyond ten days, model execution was terminated.

2.4.2   Temperature

Two estimation procedures for missing air temperature values were used. Generally, reported temperature values were read from thermometers. Some stations had a hygrothermograph as a backup instrument for humidity and temperature. For the stations maintained by IRRI, readings from the backup instrument are reported for when the thermometer failed.

With only one or two consecutive days of missing temperature data, linear interpolation was performed. Iterative interpolation was performed for months with two to ten days of missing temperature values using the decade mean as initial estimates of the missing values. Values were flagged as −99 whenever more than ten days in a month had missing values.

2.4.3   Vapor pressure

Data centers differed in their measurement of water vapor concentration in the air. Some data centers reported relative humidity, while other centers reported actual vapor pressure. For standardization purposes, actual early morning vapor pressure values were stored. Since relative humidity is the ratio of actual and saturated vapor pressure, and saturated vapor pressure is a function of temperature, actual vapor pressure data were derived from reported relative humidity percentages using mean daily temperatures. Also, the frequency of observations within a day varied between centers. Stations with more than one daily air water concentration measurement report daily means. Others simply report the early morning observation. For Los Baños, the difference between the daily mean and early morning reading of vapor pressure falls below 0.5 kPa.

2.4.4   Other weather elements

If a given month had less than ten days of missing wind velocity values, missing values were replaced with long-term monthly means. If there were more than ten days of missing values in a month, the values were coded as −99.

Data were gathered for only a few weather stations in India. Some stations had only two years of daily weather data (see Table 2.1). For such stations, daily weather data were generated for ten years using SIMWTH (Supit, 1986).

2.4.5   Potential evapotranspiration

With values of solar radiation, air temperature, vapor pressure, and wind speed, the potential evapotranspiration (PET) of a 0.2 m grass sward was estimated. We prefer PET over the Class A evaporation measurements recorded at some stations, as PET agrees more closely with real crop situations, and the Class A pan evaporation is disturbed by advective heat conducted through the walls and the bottom of the pan. The standard procedure provided by the statistical package INSTAT (Stern *et al.*, 1990) was used to derive PET.

2.5   Weather Input Format for Crop Models

After validation, the weather file was saved in FSE (Fortran Simulation Environment; van Kraalingen *et al.*, 1991) format, which was read by the crop simulation models used in the study (see Table 2.3 for an example). The first letters of the weather file indicated the name of the country, the following digit indicated the station number in the country, and the file extension represented the last three digits of the year. Thus, for example, the data from Philippines station number 1 (set number for IRRI Wetland Agromet Station) in 1993 were contained in the weather file PHIL1.993.

2.6   Summary Statistics of Stations Covered by the Rice–Weather Database

The summary statistics of the stations covered by the database (Table 2.1) are presented in Table 2.4. Table 2.5 gives the summary statistics of the stations used in Chapter 7 for the main growing season only. Figure 2.1 shows the distribution of the rice weather stations.

Table 2.3.   Example of part of the contents of a weather file (PHIL1.993), IRRI Wetland Agromet station, Los Baños, The Philippines.

*	Station Name :	IRWE0001		
*	Author	:	Climate Unit, IRRI −99. : nil value	
*	Source	:	International Rice Research Institute (IRRI)	
*				
*	Comments	:	This file is extracted from CLICOM database.	
*	Longitude	:	121° 15′ E Latitude : 14° 11′ N Altitude : 21.0 m	
*	Column	Daily Value		
*	1	Station number		
*	2	Year		
*	3	Day		
*	4	irradiance		$KJ\ m^{-2}\ d^{-1}$
*	5	min temperature		°C
*	6	max temperature		°C
*	7	vapor pressure		kPa
*	8	mean wind speed		$m\ s^{-1}$
*	9	precipitation		$mm\ d^{-1}$
*				

Continued

Table 2.3. *Continued*

121.25	14.18	21.0	0.00	0.00			
1 1993	1	19582	20.5	29.2	2.44	1.7	0.0
1 1993	2	19438	22.4	29.1	2.36	1.8	0.0
1 1993	3	10547	20.5	27.5	2.46	1.7	0.0
1 1993	4	23434	22.0	29.6	2.52	2.2	0.0
1 1993	5	19582	18.7	29.5	2.16	2.0	2.0
1 1993	6	16091	20.6	28.5	2.37	2.1	0.0
1 1993	7	10043	21.2	27.6	2.56	2.0	1.2
1 1993	8	16667	22.2	28.6	2.54	2.3	0.0
1 1993	9	10043	20.0	26.5	2.56	1.7	2.0
1 1993	10	23794	21.5	29.0	2.55	2.5	0.0
1 1993	11	9431	22.0	26.5	2.34	1.9	0.0
1 1993	12	12815	23.0	28.0	2.52	2.3	0.1
1 1993	13	18611	21.7	29.5	2.47	2.4	1.2
1 1993	14	20302	22.4	29.6	2.65	1.9	0.6
1 1993	15	22210	22.0	30.5	2.71	1.3	0.0
1 1993	16	15227	23.7	30.0	2.80	1.7	0.0
1 1993	17	9575	24.0	27.0	2.57	2.1	0.0
1 1993	18	13175	23.5	28.2	2.56	2.0	0.0
1 1993	19	15227	23.7	28.5	2.57	2.3	0.0
1 1993	20	8711	23.0	27.6	2.48	2.1	0.1
1 1993	21	11483	22.5	27.2	2.57	1.9	0.5
1 1993	22	19222	22.5	28.5	2.43	2.2	0.0
1 1993	23	24874	19.5	30.2	2.29	1.5	0.0
1 1993	24	21490	22.0	29.4	2.36	2.1	0.0
1 1993	25	21022	21.9	28.6	2.02	2.0	7.1
1 1993	26	17639	20.5	28.0	2.05	1.5	0.0
1 1993	27	19438	19.5	27.1	2.17	1.4	0.0
1 1993	28	6084	20.5	23.5	2.27	2.6	4.8
1 1993	29	11375	20.1	24.5	2.13	2.3	0.2
1 1993	30	10403	21.2	25.2	2.34	2.3	0.4
1 1993	31	9575	21.1	26.0	2.33	2.3	0.0
1 1993	32	11987	22.6	26.5	2.23	3.8	0.3
1 1993	33	12203	21.9	25.6	1.97	3.1	0.0
1 1993	34	11015	21.1	25.6	2.06	2.4	0.0
1 1993	35	11015	21.2	26.1	2.29	2.0	0.0
1 1993	36	14507	21.8	27.5	2.40	2.3	0.1
1 1993	37	20518	22.0	28.3	2.41	2.3	0.0

Table 2.4. Long-term summary of the stations described in Table 2.1.

Station	Rainfall (mm)	Met rainy days	Crop rainy days	Pet (mm)	Solar radn (MJ m⁻² d⁻¹)	Max. temp. (°C)	Min. temp. (°C)	Av. temp. (°C)	Highest temp. (°C)	Day of highest	Lowest temp. (°C)	Day of lowest	Av. hum. (%)	Lowest hum. (%)	Day of lowest	Vapor press. (kPa)	Vap. press. def. (kPa)	Wind speed (m s⁻¹)
Bangladesh																		
Jessore	2066	106	74	1197	16.7	31.4	20.9	26.2	41.5	Apr	5.9	Jan	93	43	Mar	3.26	0.26	3.3
Joydebpur	2299	102	70	1471	16.9	30.7	20.7	25.7	39.0	May	7.3	Dec	89	54	Mar	2.74	0.37	m
Khulna	2394	115	80	1157	17.4	31.0	21.4	26.2	40.1	May	7.4	Jan	96	34	Apr	3.37	0.15	2.1
Rangpur	3055	101	72	1268	16.9	29.5	19.9	24.7	39.6	May	5.9	Feb	77	42	Mar	2.52	0.07	2.1
China																		
Beijing	541	53	23	841	14.7	17.8	7.1	12.5	38.1	Jun	-15.5	Jan	55	6	Apr	1.10	0.70	2.4
Chansha	1360	125	65	676	12.0	21.4	13.8	17.5	39.2	Aug	-6.4	Jan	78	28	Jul	1.75	0.56	2.1
Chendu	887	97	36	513	9.7	19.9	12.8	16.3	35.0	Aug	-3.9	Jan	83	52	May	1.71	0.35	1.1
Fuzhou	1290	117	61	753	12.2	24.2	16.6	20.4	39.9	Jul	0.5	Feb	76	28	Jan	2.01	0.62	2.4
Guangzhou	1742	121	69	710	12.4	26.2	18.8	22.5	38.1	Jul	2.3	Jan	77	27	Dec	2.30	0.62	1.7
Guiyang	1030	119	47	627	10.7	19.6	12.2	15.9	35.1	Jul	-13.3	Jan	76	24	Jan	1.52	0.49	1.9
Hangzhou	1470	121	73	651	12.8	20.4	12.9	16.6	39.1	Jul	-7.3	Jan	77	20	Mar	1.72	0.47	2.3
Nanjing	1043	97	47	648	13.1	19.9	11.1	15.5	38.5	Jul	-11.0	Jan	77	30	Jan	1.62	0.44	2.5
Shenyang	718	70	34	733	13.5	13.9	3.6	8.8	35.2	Aug	-28.8	Jan	61	17	Mar	0.99	0.48	2.8
Wuhan	1367	88	47	632	13.4	20.7	12.7	16.7	38.5	Jul	-12.8	Jan	78	29	Feb	1.73	0.48	2.0
Jinhua	1434	126	69	1137	14.0	22.0	13.6	17.8	40.7	Aug	-9.6	Jan	77	27	Feb	1.79	0.56	6.0
India																		
Aduthurai	1116	63	40	2022	19.2	32.7	23.7	28.3	41.1	May	13.0	Mar	82	37	Sep	3.19	0.72	2.1
Bijapur	662	64	26	2027	18.9	33.2	20.8	27.0	43.6	May	10.2	Dec	m	m	m	m	m	m
Coimbatore	651	52	26	1911	18.2	31.6	21.0	26.3	38.5	May	-2.0	Oct	78	31	Feb	2.40	0.17	m
Cuttack	1569	56	36	1675	17.0	31.8	22.1	27.0	41.4	Jun	10.1	Jan	80	12	Feb	2.78	0.90	1.5

Location																		
Hyderabad	706	53	36	1270	18.4	32.1	19.6	25.9	43.1	May	6.5	Nov	86	51	May	2.35	0.42	2.1
Kapurthala	767	15	10	1477	17.1	29.8	15.2	22.5	44.8	May	-1.5	Jan	88	48	May	2.11	0.39	1.1
Madurai	824	47	27	2133	20.7	34.0	23.7	28.8	41.0	Jun	15.0	Feb	82	22	Jul	3.26	0.79	1.4
Patancheru	898	78	40	2122	18.8	31.8	20.0	25.9	43.2	May	5.4	Jan	55	12	May	1.83	1.62	2.9
Pattambi	2744	101	72	1643	17.4	31.6	22.9	27.2	39.3	Mar	15.5	Dec	93	63	Feb	2.85	0.22	0.5
Indonesia																		
Bandung	2173	147	84	1188	15.1	28.0	17.9	23.0	32.0	Oct	12.4	Jul	79	37	May	2.23	0.60	0.7
Ciledug	2312	154	98	1611	16.8	31.7	22.7	27.2	36.6	Oct	17.8	Sep	82	57	Aug	3.00	0.65	0.8
Cimanggu	4738	158	134	1112	14.5	30.5	22.8	26.6	33.5	Mar	19.5	Jun	79	62	Jun	2.80	0.38	1.0
Cipanas	2756	135	114	1050	m	24.0	17.1	20.5	27.8	Oct	12.6	Jul	83	66	Oct	2.03	0.41	m
Maros	3252	141	98	1718	17.7	31.0	22.3	26.6	36.7	Nov	15.5	Oct	91	59	Aug	3.22	0.31	m
Muara	4215	122	97	1395	14.0	31.0	21.4	26.2	33.5	Oct	17.5	Mar	69	50	Oct	2.36	0.91	0.9
Pacet	3232	199	139	979	12.2	24.8	16.8	20.8	28.6	Sep	12.5	Jul	85	15	Oct	2.11	0.38	m
Japan																		
Akita	1717	170	96	739	11.6	15.1	7.6	11.3	36.1	Aug	-11.8	Jan	73	31	Jun	1.16	0.35	0.9
Hiroshima	1666	96	68	906	13.1	19.7	11.6	15.6	37.0	Aug	-7.5	Feb	71	32	Apr	1.46	0.47	0.7
Kochi	2346	112	81	927	14.0	21.5	12.0	16.7	37.3	Aug	-6.1	Feb	68	26	Apr	1.49	0.53	0.5
Maebashi	1205	103	59	798	12.5	19.0	9.8	14.4	39.1	Jul	-9.0	Feb	65	25	Mar	1.26	0.51	0.7
Miyazaki	2497	118	81	962	13.7	21.9	12.9	17.4	37.6	Jul	-5.9	Jan	74	37	Apr	1.67	0.45	0.6
Nagoya	1534	106	70	890	12.9	20.1	11.4	15.7	38.2	Aug	-6.2	Feb	67	32	Apr	1.39	0.54	0.7
Sapporo	1078	140	65	629	11.9	12.5	4.7	8.6	34.8	Aug	-16.8	Jan	70	30	Apr	0.95	0.33	0.5
Sendai	1247	106	58	710	12.1	15.9	8.4	12.2	36.0	Sep	-8.6	Feb	71	30	Apr	1.19	0.36	0.8
Toyama	2231	176	117	794	11.7	17.7	9.7	13.7	37.1	Aug	-8.3	Jul	78	36	May	1.38	0.35	0.7
Malaysia																		
Kemubu	2836	157	105	1505	18.3	30.8	23.1	26.9	39.0	Apr	17.5	Feb	76	65	Jun	2.75	0.64	2.4
Tanjung Karang	1835	136	74	1484	18.7	31.3	23.1	27.2	36.4	Mar	19.4	Jun	82	56	Feb	2.99	0.65	m
Telok Chengai	2243	142	91	1543	19.2	31.7	23.4	27.6	37.2	Apr	18.1	Jan	82	56	Feb	3.05	0.67	m

Continued

Table 2.4. *Continued*

Station	Rainfall (mm)	Met rainy days	Crop rainy days	Pet (mm)	Solar radn (MJ m⁻² d⁻¹)	Max. temp. (°C)	Min. temp. (°C)	Av. temp. (°C)	Highest temp. (°C)	Day of highest	Lowest temp. (°C)	Day of lowest	Av. hum. (%)	Lowest hum. (%)	Day of lowest	Vapor press. (kPa)	Vap. press. def. (kPa)	Wind speed (m s⁻¹)
Myanmar																		
Myananda	662	51	33	m	21.5	33.3	20.9	27.1	42.5	May	8.0	Jan	m	m	m	m	m	m
Pyay	1113	93	59	m	27.7	33.6	22.3	28.0	43.7	Mar	8.0	Jan	m	m	m	m	m	m
Wakema	2321	145	110	m	21.2	32.0	21.2	26.6	39.5	Apr	9.5	Jan	m	m	m	m	m	m
Yezin	1195	98	57	1844	18.2	32.6	21.2	26.9	41.0	Apr	9.0	Jan	92	58	Mar	2.50	0.07	1.9
Philippines																		
Albay	4995	190	129	1566	16.2	30.9	22.4	26.7	40.0	Apr	15.2	Jun	90	59	Sep	3.35	0.37	1.2
Banaue	3608	208	147	1380	15.1	25.3	17.3	21.3	31.9	Jun	8.5	Jan	89	52	Mar	2.21	0.28	0.5
Betinan	2891	186	125	1652	16.0	31.0	22.2	26.6	39.5	May	18.0	Jan	92	61	Jun	2.91	0.33	0.8
Butuan	2070	164	91	1681	16.6	32.1	22.5	27.3	37.8	May	17.5	Feb	88	73	Mar	3.15	0.43	1.3
Cavinti	4196	244	151	1517	16.4	28.1	20.7	24.4	37.1	Aug	10.0	Nov	92	77	Dec	2.81	0.25	2.0
Claveria	1842	157	90	1677	18.9	28.6	21.3	25.0	33.0	Sep	16.8	Dec	84	56	Feb	2.62	0.52	1.1
CLSU	1944	111	72	1928	18.9	31.8	22.4	27.1	38.5	May	15.1	Dec	85	17	Jun	2.93	0.55	2.4
CSAC	2162	155	82	1528	m	31.7	22.7	27.2	38.0	Jun	15.2	Feb	88	66	Apr	3.13	0.44	1.2
DAET	3369	187	117	1693	m	30.8	23.6	27.2	37.0	May	16.3	Sep	87	70	Mar	3.07	0.46	2.6
Guimba	1656	102	60	1986	21.0	32.6	22.0	27.3	39.5	Jan	12.5	Dec	89	62	Apr	2.90	0.36	1.6
H. Luisita	1842	78	48	1836	18.3	32.9	21.4	27.1	39.9	Jun	12.9	Dec	82	61	Apr	3.24	0.13	2.1
Iloilo	2080	133	76	1671	m	31.1	24.0	27.6	37.8	May	16.5	Jan	86	64	Apr	3.13	0.51	2.0
IRRI Dryland	2091	146	79	1786	16.1	31.5	23.0	27.3	38.2	May	15.2	Feb	88	71	May	3.02	0.44	1.6
IRRI Wetland	2027	150	79	1741	16.1	30.6	23.2	26.8	36.0	May	16.8	Feb	88	70	May	3.05	0.44	1.5
La Granja	2589	155	102	1539	13.7	32.3	22.0	27.1	39.5	Apr	15.5	Feb	86	57	Apr	3.05	0.50	1.8
MMSU	1876	84	54	1765	18.4	32.1	21.9	27.0	37.3	May	10.4	Jan	87	50	Oct	2.95	0.46	0.9

MSAC	3641	151	100	1255	16.2	23.7	14.5	19.1	29.0	Apr	5.0	Dec	88	55	Mar	1.97	0.28	1.9
Munoz	1581	100	68	m	20.8	31.4	23.0	27.2	37.2	May	15.0	Dec	89	64	Dec	2.90	0.37	1.9
NARS	2470	131	81	1665	17.1	31.9	22.0	27.0	39.3	May	14.0	Dec	86	54	Jan	3.01	0.50	1.1
PNAC	1636	120	66	1772	18.4	31.2	23.2	27.2	36.5	Sep	16.2	Jan	90	46	Nov	3.23	0.37	1.1
Solana	1422	106	58	1755	18.5	31.7	22.5	27.1	39.5	May	13.0	Feb	92	65	May	2.88	0.27	0.9
UPLB	2035	152	81	1581	16.9	31.6	22.6	27.1	38.3	May	15.6	Jan	87	46	May	3.05	0.48	1.2
Ves	2200	142	84	1809	18.3	31.4	23.5	27.5	38.1	May	16.1	Dec	91	56	May	3.26	0.35	1.4
South Korea																	
Cheolweon	1537	86	45	806	13.2	15.4	4.3	9.8	35.2	Jul	-32.1	Jan	77	16	May	1.19	0.33	m
Ch'iigok	1298	63	37	653	16.6	19.2	6.6	12.9	39.2	Jul	-16.7	Jan	69	23	May	1.28	0.47	1.5
Chinju	1587	83	51	606	13.9	19.1	7.5	13.3	37.0	Aug	-15.9	Jan	73	21	Apr	1.38	0.41	1.7
Chonju	1453	100	55	906	12.9	18.3	8.1	13.2	36.4	Jul	-14.9	Jan	72	33	Apr	1.34	0.48	1.3
Chongju	1738	81	47	879	14.1	19.7	8.4	14.1	37.8	Aug	-22.1	Jan	72	3	Mar	1.43	0.50	1.8
Ch'unch'on	1388	82	45	561	12.8	16.9	5.5	11.2	36.5	Jul	-25.6	Jan	72	33	Sep	1.22	0.43	1.4
Kangnung	1569	81	47	986	12.8	17.3	8.7	13.0	37.9	Jul	-15.7	Feb	63	16	Jun	1.17	0.57	2.4
Kwangju	1485	100	54	962	13.3	18.7	8.8	13.7	36.9	Aug	-12.6	Feb	73	23	Aug	1.40	0.46	2.0
Milyang	1262	82	50	654	14.7	19.3	7.1	13.2	38.1	Aug	-17.5	Jan	68	21	Feb	1.28	0.51	1.5
Suweon	1341	83	45	822	12.7	16.8	6.3	11.6	35.4	Aug	-24.8	Jan	73	25	Apr	1.25	0.41	1.5
Taejon	1691	94	53	595	13.3	17.8	7.2	12.5	35.8	Aug	-16.8	Jan	72	24	Apr	1.31	0.44	1.6
Taiwan																	
Pingtung	2108	89	56	1386	13.1	29.2	20.4	24.8	37.3	Jul	5.8	Dec	83	54	Feb	2.65	0.37	1.9
Thailand																	
Chiang Mai	1140	94	54	1424	18.5	32.0	20.6	26.3	41.3	Apr	7.1	Dec	71	34	Mar	2.46	1.04	m
Khon Kaen	1228	90	54	1554	18.8	32.7	22.0	27.3	42.6	Apr	5.6	Dec	70	34	Mar	2.62	1.11	m
Nakhon S.	1104	89	51	1689	19.0	34.0	23.2	28.6	42.6	May	8.2	Dec	70	35	Mar	2.80	1.21	m
Ubon Ratchathani	1652	98	62	1583	19.1	32.5	21.8	27.2	41.3	Apr	8.5	Dec	72	42	Mar	2.65	1.04	m
Surat Thani	1568	130	70	m	18.6	32.0	22.3	27.2	38.2	Apr	14.3	Jan	81	63	Mar	2.97	0.68	m

m = missing data.

Table 2.5. Mean temperature and solar radiation during the main growing season for each of the weather stations used in the study.

Country	Site name	Growing season averages	
		Temperature (°C)	Solar radiation (MJ m^{-2} d^{-1})
Bangladesh	Khulna	29.1	18.0
	Jessore	29.4	18.0
	Rangpur	28.3	18.9
	Joydebpur	28.9	17.4
China	Beijing	22.6	17.1
	Chansha	27.5	17.8
	Chendu	24.5	13.7
	Fuzhou	26.8	16.2
	Guangzhou	27.6	15.1
	Guiyang	23.4	15.0
	Hangzhou	26.4	17.3
	Nanjing	25.7	17.1
	Wuhan	26.9	18.8
	Shenyang	23.2	17.6
India	Aduthurai	29.3	19.2
	Coimbatore	26.3	18.7
	Cuttack	27.1	15.7
	Kapurthala	24.6	16.4
	Patancheru	24.5	17.7
	Bijapur	26.4	15.6
	Madurai	30.0	21.0
	Pattambi	26.2	15.1
Indonesia	Bandung	23.0	12.1
	Ciledug	27.2	13.5
	Cimanggu	26.4	13.5
	Maros	26.7	16.3
	Muara	26.2	13.4
	Pacet	20.8	10.9
Japan	Akita	21.3	16.5
	Hiroshima	24.4	16.3
	Kochi	25.4	15.8
	Miyazaki	26.0	16.0
	Maebashi	23.9	12.8
	Nagoya	24.9	14.9
	Sapporo	18.4	15.4
	Sendai	20.5	14.7
	Toyama	21.6	15.0

Continued

Table 2.5. *Continued*

		Growing season averages	
Country	Site name	Temperature (°C)	Solar radiation (MJ m^{-2} d^{-1})
Malaysia	Kemubu	26.9	16.6
	Telok Chengai	27.1	17.2
	Tanjung Karang	26.9	17.5
Myanmar	Myananda	29.7	20.8
	Pyay	28.5	24.2
	Wakema	26.9	14.6
	Yezin	27.8	17.9
Philippines	Albay	27.6	16.7
	Butuan	27.7	17.4
	CLSU	27.2	18.1
	La Granja	27.1	13.5
	MMSU	27.7	17.3
	PNAC	27.3	17.1
	UPLB	27.5	16.2
	IRRI Wetland	27.5	15.6
South Korea	Chongju	22.0	15.2
	Ch'ilgok	21.4	19.5
	Chinju	22.2	15.5
	Ch'unch'on	21.4	15.2
	Daejun	21.5	15.0
	Kangnung	20.2	13.9
	Kwanju	23.3	15.8
	Chonju	23.0	15.0
	Milyang	22.1	16.8
	Suweon	21.5	14.4
	Cheolweon	21.1	16.2
Taiwan	Pingtung	22.1	13.0
Thailand	Chiang Mai	27.6	17.5
	Khon Kaen	27.8	17.8
	Ubon Ratchathani	27.4	17.8

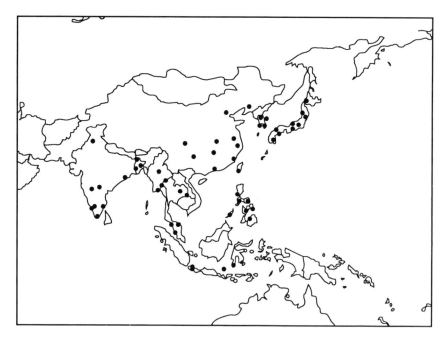

Fig. 2.1.   Locations of the weather stations used in the study.

2.7   Acknowledgments

We gratefully acknowledge the help of the following research institutions in making their weather data available for our study:

Bangladesh Rice Research Institute, Joydebpur, Bangladesh
China National Rice Research Institute, Hangzhou, China
Crop Experimental Station, Suweon, South Korea
International Rice Research Institute, Los Baños, The Philippines
Kyoto University, Kyoto, Japan
Oregon State University, Oregon, USA
Tamil Nadu Rice Research Institute, Tamil Nadu, Aduthurai, India
Universiti Pertanian Malaysia, Kuala Lumpur, Malaysia

We are also grateful to L. R. Oldeman, D. V. Seshu, F. B. Cady, and their colleagues in the IRRI Rice–Weather Project, for making their weather data available to form the basis of our current database.

The Rice Model ORYZA1 and Its Testing   3

M.J. KROPFF[1], R.B. MATTHEWS[1,2], H.H. VAN LAAR[3], AND
H.F.M. TEN BERGE[2]

with contributions from J.C. Shin[4], S. Mohandass[5], S. Singh[6], Zhu
Defeng[7], Moon Hee Lee[4], A. Elings[2], and B.A.M. Bouman[2]

[1]International Rice Research Institute, P.O. Box 933, 1099 Manila, The
Philippines; [2]Research Institute for Agrobiology and Soil Fertility,
Bornsesteeg 65, 6700 AA Wageningen, The Netherlands; [3]Department of
Theoretical Production Ecology, Wageningen Agricultural University, P.O.
Box 430, 6700 AK Wageningen, The Netherlands; [4]Crop Experiment
Station, Seodundong 209, Suweon, South Korea; [5]Tamil Nadu Rice
Research Institute, Aduthurai, Tamil Nadu 612101, India; [6]Universiti
Pertanian Malaysia, 43400 UPM Serdang, Selangor, Malaysia; [7]China
National Rice Research Institute, 171 Tiyuchang Road, Hangzhou,
310006 Zhejiang Province, China

3.1   Introduction

The ORYZA1 model (v2.11) describing the potential production of rice was
modified from the SUCROS model (Spitters *et al.*, 1989; van Laar *et al.*, 1992),
the MACROS module L1D (Penning de Vries *et al.*, 1989), the INTERCOM
model (Kropff & van Laar, 1993), and the GUMCAS cassava model (Matthews
& Hunt, 1994). Recently obtained data sets, in which the potential yields of
modern rice varieties in flooded systems were obtained by careful nitrogen (N)
management and intensive crop protection, were used for its development and
calibration (Kropff *et al.*, 1993). A model containing detail of processes at the
leaf level was chosen, as much of the knowledge of CO_2 and temperature effects
on growth processes is available at this level of organization.

3.2   General Structure of the Model ORYZA1

The general structure of the model is presented in Fig. 3.1. Under favorable
growth conditions, light, temperature, and the varietal characteristics for
phenological, morphological, and physiological processes are the main factors

determining the growth rate of the crop on a specific day. The model follows a daily calculation scheme for the rates of dry matter production of the plant organs, the rate of leaf area development, and the rate of phenological development. By integrating these rates over time, dry matter production of the crop is simulated throughout the growing season.

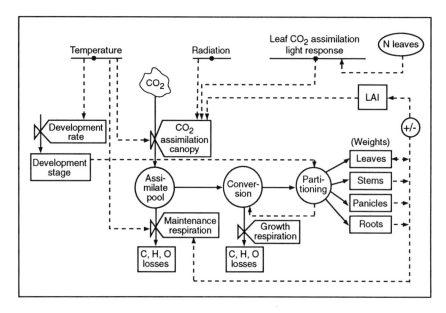

Fig. 3.1. A schematic representation of the model ORYZA1. Boxes are state variables, valves are rate variables, and circles are intermediate variables. Solid lines are flows of material and dashed lines are flows of information.

The total daily rate of canopy CO_2 assimilation is calculated from the daily incoming radiation, temperature, and the leaf area index (LAI). The model contains a set of subroutines that calculate the daily rate by integrating instantaneous rates of leaf CO_2 assimilation. The calculation is based on an assumed sinusoidal time course of solar radiation over the day and the exponential light profile within the canopy. On the basis of the photosynthesis characteristics of single leaves, which depend upon the N concentration, the photosynthesis profile in the canopy is obtained. The effect of the ambient CO_2 concentration on the photosynthetic parameters was also included in this version of the model. Integration over the LAI of the canopy and over the day gives the daily CO_2 assimilation rate. After subtraction of respiration requirements, the net daily growth rate (kg ha^{-1} d^{-1} dry matter) is obtained. The dry matter produced is partitioned among the various plant organs.

Phenological development rate is tracked as a function of daily average ambient temperature. Photoperiod may also affect the rate of development. Before the canopy is fully closed, the increment in leaf area is calculated from daily average temperature, because carbohydrate production does not usually limit leaf expansion. When the canopy closes, the increase in leaf area is obtained from the increase in leaf weight. Integration of daily growth rates of the organs and leaf area results in dry weight increment during the growing season. A simple procedure was incorporated into the model to simulate sink limitation as a result of spikelet sterility at high or low temperatures.

Input requirements of the model are: geographical latitude, daily weather data (solar radiation, minimum and maximum temperature), plant density, date of crop emergence and transplanting, and parameter values that describe the morpho-physiological characteristics of rice. The time step of integration is one day.

3.3   Phenological Development of the Crop

3.3.1   General

The developmental stage of a plant defines its physiological age and is characterized by the formation of the various organs and their appearance. The most important phenological events are emergence, panicle initiation, flowering and maturity. As many physiological and morphological processes change with the phenological stage of the plant, accurate quantification of phenological development is essential in any simulation model for plant growth. Temperature and photoperiod are the main driving forces of developmental changes.

The concept of a developmental day is used to describe the integrated effects of both temperature and photoperiod on phenology. A development day (Dd) is defined as the advancement in the phenological age of the crop over one chronological day at the optimum temperature and photoperiod, where other factors such as nutrients, water, and pests and diseases do not limit growth (Matthews & Hunt, 1994). The number of developmental days between certain events, such as panicle initiation to flowering, is assumed to be constant for a given variety, regardless of the prevailing temperature or photoperiod. However, suboptimal temperatures or photoperiods will result in more chronological days being required to reach the given stage of development.

In the modified version of the model ORYZA1, the life cycle of the rice crop is divided into four main phenological phases:

1. The basic vegetative phase (BVP), from sowing to the start of the photoperiod-sensitive phase.
2. Photoperiod-sensitive phase (PSP), during which the rate of development is influenced by daylength, lasting until panicle initiation (PI).

3. Panicle formation phase (PFP), starting at the switch from the vegetative to the reproductive state at PI and lasting until 50% flowering.
4. Grain-filling phase (GFP), from 50% flowering to physiological maturity.

It is assumed in the model that for a given variety, each of these phases requires a constant number of developmental days for its completion, and that the rate at which developmental days accumulate is influenced by temperature in all four phases, and additionally, by photoperiod in the photoperiod-sensitive phase. Typically, the duration of each phase is 20, 15, 25, and 25 Dd, respectively, for a variety such as IR64. Differences between varieties in the total crop duration is usually due to differences in the duration of the BVP rather than the other phases (Vergara & Chang, 1985).

3.3.2   Effects of temperature on phenological development

It has been observed in many crops that the rate of development (i.e. the reciprocal of the duration in days it takes to complete a certain phenological event, such as flowering) is linearly related to the daily mean temperature above a base temperature up to an optimum temperature, beyond which the rate decreases, again linearly, until a maximum temperature is reached (e.g. Kiniry *et al.*, 1991). For temperatures below the base temperature or above the maximum temperature, the rate of development is zero. That the development rate indeed decreases at temperatures above the optimum was shown by P.A. Manalo, K.T. Ingram, R.R. Pamplona, and A.O. Egeh (unpublished). In four varieties (IR28, IR36, IR64, ITA186) they found a longer period to flowering at a mean temperature of 33°C in comparison to 29°C. However, the magnitude of the simulated increase in duration above 32°C has not been fully tested because of lack of data.

Three 'cardinal' temperatures can therefore be identified: the base temperature (T_{base}), the optimum temperature (T_{opt}), and the maximum temperature (T_{high}). For rice, these values are typically 8°C, 30°C and 42°C, respectively (Gao *et al.*, 1992). This 'bilinear' response is generally observed only when the daily temperatures are constant (e.g. in a controlled environment); if the temperature fluctuates between a minimum and a maximum value, as is the case in normal field experiments, the response becomes more curvilinear, particularly near each cardinal temperature. The linear and curvilinear responses are shown in Fig. 3.2.

Although this curvilinear response to daily mean temperature can be described by complex exponential equations (e.g. Gao *et al.*, 1992), the simpler approach used by Matthews & Hunt (1994) in their cassava model was used in this version of the ORYZA1 model. In this approach, it is assumed that the response of development rate to temperature over short time periods, such as one hour, is described by the bilinear model, and that the response to daily mean temperature is achieved by superimposing on to this model a

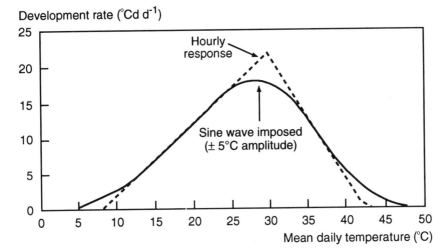

Fig. 3.2. The response functions of phenological development rate to temperature as used in ORYZA1 version 1.22. Simulations with $T_{base} = 8°C$, $T_{opt} = 30°C$, and $T_{max} = 42.5°C$.

temperature response approximated by a sine function alternating between the daily minimum (T_{min}) and maximum (T_{max}) temperatures (Fig. 3.2).

In the model, hourly temperature (T_d) is calculated from T_{min} and T_{max} according to the relation:

$$T_d = (T_{min} + T_{max})/2 + (T_{max} - T_{min}) \cos[0.2618 (h - 14)]/2 \qquad (3.1)$$

where h is the hour of the day. Hourly increments in developmental age (HD, Dd d^{-1}) are calculated according to:

$$
\begin{array}{lll}
T_d \leq T_{base}, T_d \geq T_{high} & : & HD = 0 \\
T_{base} < T_d \leq T_{opt} & : & HD = (T_d - T_{base}) / (T_{opt} - T_{base}) \\
T_{opt} < T_d < T_{high} & : & HD = (T_{high} - T_d) / (T_{high} - T_{opt}) \qquad (3.2)
\end{array}
$$

The daily increment in developmental time ($f[T_{min}, T_{max}]$, Dd d^{-1}) is then calculated as:

$$f(T_{min}, T_{max}) = \sum_{h=1}^{24} (HD/24) \qquad (3.3)$$

3.3.3   Effects of photoperiod on phenological development

The effect of photoperiod (ϕ, h) is simulated using a modification of the function for a short-day plant described by Major & Kiniry (1991), expressing the response as a rate rather than a duration. The function (g[φ]) can be characterized by two parameters, the maximum optimum photoperiod (ϕ_o, h) and photoperiod sensitivity (S, h^{-1}):

$$\text{for} \quad \phi \leq \phi_o: g(\phi) = 1$$
$$\phi > \phi_o: g(\phi) = 1 - S(\phi_o - \phi), g(\phi) \geq 0 \quad (3.4)$$

3.3.4 Effects of transplanting shock on phenological development

In transplanted rice, the situation becomes more complicated because of the transplanting shock, which causes a delay in phenological development. In experiments, it was found that the delay in phenological development (expressed in developmental age) is a function of the developmental age of the seedlings that are transplanted. The way in which this is incorporated in the model is illustrated in Fig. 3.3. As the model proceeds, the developmental age of the crop is tracked. At transplanting, the duration of the transplanting shock period is determined from the developmental age of the seedlings using a dimensionless coefficient relating the two variables. If the date of sowing equals the date of transplanting (i.e. in direct-seeded rice), it is assumed that the transplanting shock period is zero.

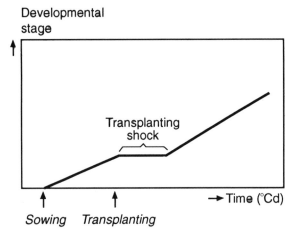

Fig. 3.3. Procedure for simulation of transplanting shock effect on phenological development. After transplanting the developmental rate is zero for a period expressed in degree-days (°Cd).

3.4 Daily Rate of Gross CO_2 Assimilation of the Canopy

The calculation procedure for the daily rate of crop CO_2 assimilation, developed by Spitters *et al.* (1986), is schematically represented in Fig. 3.4. Measured or

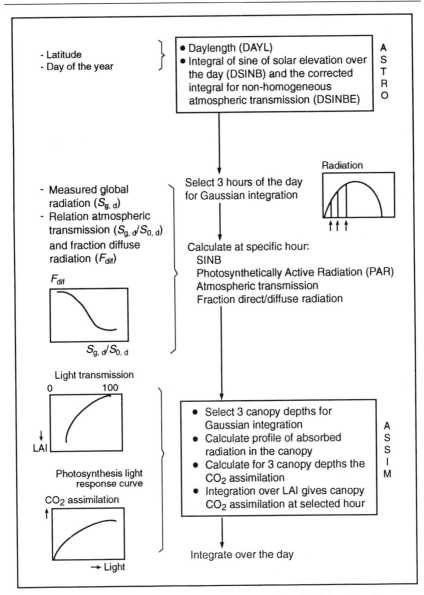

Fig. 3.4. Schematic representation of the calculation procedure for daily rates of canopy CO_2 assimilation in the subroutine TOTASS, which calls the subroutines ASTRO and ASSIM.

estimated daily total solar irradiation (wavelength 300–3000 nm) is input into the model. Only half of this incoming solar radiation is photosynthetically active radiation (*PAR*, wavelength 400–700 nm). This visible fraction, generally called 'light', is used in the calculation procedure of the CO_2 assimilation rate of the canopy.

A distinction is made between diffuse skylight with various angles of incidence, and direct sunlight with an angle of incidence equal to the solar angle. It is important to distinguish these fluxes because of the large difference in illumination intensity between shaded leaves (receiving only diffuse radiation) and sunlit leaves (receiving both direct and diffuse radiation) and the non-linear CO_2 assimilation–light response of single leaves. The proportion of diffuse light in the total incident light flux depends on the status of the atmosphere (e.g. cloudiness, concentration of aerosols), and is calculated from the atmospheric transmission using an empirical function based on data from different meteorological stations from a wide range of latitudes and longitudes (Spitters *et al.*, 1986). The atmospheric transmission is the ratio between actual irradiance *PAR* (measured in J m^{-2} s^{-1}) and the quantity that would have reached the earth's surface in the absence of an atmosphere. The fluxes of direct and diffuse *PAR* are calculated from the diffuse radiation fraction.

Radiation fluxes attenuate exponentially within a canopy with increasing leaf area from the top downwards:

$$I_L = (1 - r) I_0 \exp(-kL) \qquad (3.5)$$

where I_L is the net *PAR* flux at depth L in the canopy (with an *LAI* of L above that point) (J m^{-2} ground s^{-1}), I_0 is the flux of visible radiation at the top of the canopy (J m^{-2} ground s^{-1}), L is the cumulative LAI (counted from the top of the canopy downwards) (m^2 leaf m^{-2} ground), r is the reflection coefficient of the canopy (dimensionless), and k is the extinction coefficient for *PAR* (dimensionless). The diffuse and the direct flux have different extinction coefficients, giving rise to different light profiles within the canopy for diffuse and direct radiation.

For a spherical leaf angle distribution (homogeneous, random), the extinction coefficient for diffuse radiation (k_{df}) is about 0.71 (Goudriaan, 1977). However, in many crops, the leaf angle distribution is not spherical. In rice, the leaves are clustered (especially in the beginning as a result of planting on hills), and have a very vertical orientation. Other leaf angle distributions can be accounted for by a procedure described by Goudriaan (1986), which calculates k_{df} on the basis of the frequency distribution of leaves with angles in three classes (0–30°, 30–60°, and 60–90°). In the model, however, this is accounted for by using a cluster factor (Cf) which is the measured k_{df}, relative to the theoretical one, for a spherical leaf angle distribution, calculated as follows:

$$Cf = k_{df} / [0.8 \sqrt{(1 - \sigma)}] \qquad (3.6)$$

k_{df} is the measured extinction coefficient under diffuse sky conditions and is an input into the model, and σ is the scattering coefficient. Values for k_{df} range from

0.4–0.7 for monocotyledons (erectophile) and 0.65–1.0 for dicotyledons (Monteith, 1969). For rice a value of 0.4 is used until the canopy closes and 0.6 for a closed canopy. This accounts for the clustering of leaves by planting on hills and the very erect stature of the leaves at early stages. For σ, a value of 0.2 is used (Goudriaan, 1977).

The light absorbed at a depth L in the canopy ($I_{a,L}$) is obtained by taking the derivative of Equation 3.5 with respect to the cumulative LAI:

$$I_{a,L} = - dI_L/dL = k (1 - r) I_0 \exp (- kL) \tag{3.7}$$

This procedure is followed for the different light components (direct and diffuse).

The CO_2 assimilation-light response of individual leaves is described by a negative exponential function characterized by the initial slope (ϵ) and an asymptote (Am) (Goudriaan, 1982):

$$A_L = A_m [1 - \exp (-\epsilon I_a / A_m)] \tag{3.8}$$

where A_L is the gross assimilation rate (kg CO_2 ha^{-1} leaf h^{-1}), A_m is the gross assimilation rate at light saturation (kg CO_2 ha^{-1} leaf h^{-1}), ϵ is the initial light use efficiency (kg CO_2 ha^{-1} leaf h^{-1} / (J m^{-2} leaf s^{-1})), and I_a is the amount of absorbed radiation (J m^{-2} leaf s^{-1}). The initial light use efficiency (ϵ) is calculated from a linear function of temperature, based on data of Ehleringer & Pearcy (1983). In C_3 species, ϵ decreases slightly with increasing temperature as the affinity of the carboxylating enzyme rubisco for oxygen (O_2) increases compared to CO_2. As these authors did not observe differences between species, such values are used for all species. This variable is assumed not to be affected by the leaf N concentration, as light is the limiting factor in the process.

The light saturated rate of leaf CO_2 assimilation (A_m), however, varies considerably, mainly as a function of leaf age and the actual environmental conditions and the environmental conditions to which the leaf has been exposed in the past. These effects are largely the result of different leaf N concentrations. Therefore, A_m in the model is calculated from the leaf N concentration, expressed per unit leaf area (van Keulen & Seligman, 1987; Penning de Vries et al., 1990; Fig. 3.5). The effect of temperature was introduced according to Penning de Vries et al. (1989), which is a general relation for C_3 plants.

The effect of CO_2 on ϵ was described by the function derived by Jansen (1990) from data by Akita (1980) and van Diepen et al. (1987):

$$\epsilon = \epsilon_{340\ ppm} \frac{1-\exp (0.00305\ CO_2 - 0.222)}{1-\exp (-0.00305 \times 340 - 0.222)} \tag{3.9}$$

This relationship gives very similar results to the theoretical relationship derived by Goudriaan & van Laar (1994). A new relationship was derived for

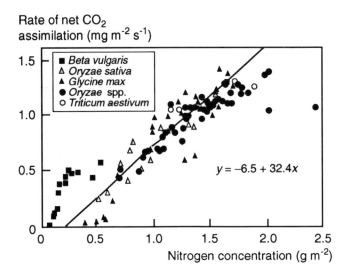

Fig. 3.5. The relationship between the maximum rate of CO_2 assimilation of single leaves and the leaf N concentration on a per area basis (g m^{-2}) (Redrawn after van Keulen & Seligman, 1987).

the effect of CO_2 on A_m from unpublished data collected at IRRI by Weerakoon, Olszyk and Ingram (unpublished) after correcting for differences in the N concentration of the leaf (Fig. 3.6):

$$A_m = A_{m,\ 340\ ppm} \langle 49.57/34.26\ \{1-\exp\left[-0.208\ (CO_2-60)/49.57\right]\}\rangle \quad (3.10)$$

From the absorbed light intensity at depth L, the assimilation rate at that specific canopy height is calculated for shaded and sunlit leaves, separately. The assimilation rate per unit leaf area at a specific height in the canopy is the sum of the assimilation rates of sunlit and shaded leaves, taking into account the proportion of sunlit and shaded leaf area at that depth in the canopy.

The maximum rate of CO_2 assimilation of a leaf (A_m, kg CO_2 ha^{-1} h^{-1}) is calculated from the N content of the leaves and the average daytime temperature. Because N content in the leaves is higher in the top leaves, these absorb most radiation, the N profile in the canopy is taken into account. Experimental measurements indicate that the profile of N in the canopy declines exponentially with *LAI* starting from the top of the canopy, with an extinction coefficient of 0.4.

A Gaussian integration procedure is used to integrate rates of leaf CO_2 assimilation over the canopy *LAI*. Three depths in the canopy are selected at which the amounts of absorbed radiation and leaf CO_2 assimilation are calculated. Total canopy assimilation is calculated as the product of the weighted

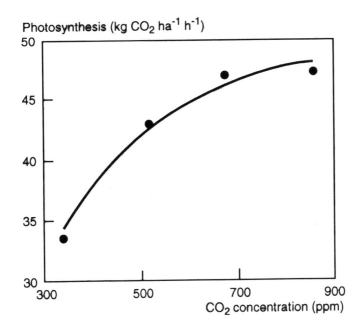

Fig. 3.6. The relationship between the maximum rate of leaf photosynthesis at 1 g N m^{-2} and external CO_2 concentration during rice growth (Data from Weerakoon, Olszyk and Ingram, IRRI/EPA, unpublished).

average of these values and total *LAI*. Daily rates of CO_2 assimilation are calculated by integrating these instantaneous rates of canopy assimilation, using a similar Gaussian integration over the time of day (Goudriaan, 1986).

In most models, for canopy CO_2 assimilation, only light absorption by leaves is accounted for, although the stems and reproductive organs may also absorb a substantial amount of radiation. In rice, for example, the stem (or sheath) area index (*SAI*) may be as high as 1.5 m^2 stem m^{-2} ground and the panicle area index (*PAI*, m^2 panicle m^{-2} ground) may be as high as 0.9 (M.J. Kropff and K.G. Cassman, IRRI, unpublished data). The model accounts for CO_2 assimilation of the stem by adding 50% of the projected green stem area (*SAI*) to the *LAI*, because sheaths are less photosynthetically active than leaves. Further information is in the section on leaf area development (Section 3.11).

3.5 Maintenance and Growth Respiration

The assimilated CO_2 is converted into carbohydrates (CH_2O) in the CO_2 assimilation process. The energy for this reduction process is provided by the absorbed light. The overall chemical reaction of this complex process is:

$$\text{6CO}_2 + \text{6H}_2\text{O} \xrightarrow{\text{light}} \text{C}_6\text{H}_{12}\text{O}_6 + \text{6O}_2$$

From this reaction it follows that for every kilogram of CO_2 taken up, 30/44 kg of CH_2O is formed; the numerical values representing the respective molecular weights of CH_2O and CO_2. Part of the carbohydrates produced in this process are respired to provide the energy for maintaining the existing biostructures, in the process known as maintenance respiration. The remaining carbohydrates are converted into structural plant dry matter. The losses in weight as a result of this conversion are termed growth respiration.

Maintenance respiration provides the energy for living organisms to maintain their biochemical and physiological status. In the reaction (which is the reverse of CO_2 assimilation, or photosynthesis), the solar energy which was fixed in the photosynthetic process in a chemical form is released in the form of ATP and NADPH:

$$\text{C}_6\text{H}_{12}\text{O}_6 + \text{6O}_2 \longrightarrow \text{6CO}_2 + \text{6H}_2\text{O} + \text{energy}$$

This process consumes roughly 15–30% of the carbohydrates produced by a crop in a growing season (Penning de Vries *et al.*, 1989), which indicates the importance of accurate quantification of this process in the model. However, the process is poorly understood at the biochemical level and simple empirical approaches are inaccurate since it is impossible to measure maintenance respiration in the way it is defined (Amthor, 1984; Penning de Vries *et al.*, 1989). The approach taken in the model is based on theoretical considerations, empirical studies and studies in which the carbon balance in the model was evaluated using crop growth and canopy CO_2 assimilation data (after Penning de Vries *et al.*, 1989).

Three components of maintenance respiration can be distinguished at the cellular level: maintenance of concentration differences of ions across membranes, maintenance of proteins and a component related to the metabolic activity of the tissue (Penning de Vries, 1975). Maintenance respiration can thus be estimated from mineral and protein concentrations and metabolic activity as presented by de Wit *et al.* (1978). In the model, we use an adapted version of the simple approach developed by Penning de Vries & van Laar (1982), in which maintenance requirements are approximately proportional to the dry weights of the plant organs to be maintained:

$$R_{m.r} = mc_{lv}\, W_{lv} + mc_{st}\, W_{st} + mc_{rt}\, W_{rt} + mc_{so}\, W_{so} \qquad (3.11)$$

where $R_{m.r}$ is the maintenance respiration rate at the reference temperature of 25°C (kg CH_2O ha^{-1} d^{-1}); W_{lv}, W_{st}, W_{rt} and W_{so} are the weights at a given time of the leaves, stems, roots, and storage organs, respectively (kg dry matter ha^{-1}); and mc_{lv}, mc_{st}, mc_{rt} and mc_{so} are the maintenance respiration coefficients for leaves, stems, roots, and storage organs, respectively (kg CH_2O kg^{-1} DM d^{-1}).

The maintenance respiration coefficients (kg CH_2O kg^{-1} dry matter d^{-1}) have different values for the different organs because of large differences in nitrogen contents. Typical values for these coefficients are 0.03 for leaves, 0.015 for stems, and 0.01 for roots (Spitters et al., 1989). For tropical crops, such as rice, lower values are used: 0.02 for the leaves and 0.01 for the other plant organs (Penning de Vries et al., 1989). In the model, for mc_{so} a coefficient of 0.003 is used, which accounts for the small fraction of active tissue in the storage organs. Maintenance respiration can also be approached by using the coefficient for stem tissue for the active part (non-stored material) only.

The effect of temperature on maintenance respiration is simulated assuming a Q_{10} of 2 (Penning de Vries et al., 1989):

$$R_m = R_{m,r} \cdot 2^{(T_{av}-T_r)/10} \qquad (3.12)$$

where R_m is the actual rate of maintenance respiration (kg CH_2O ha^{-1} d^{-1}), T_{av} is the average daily temperature (°C), and T_r is the reference temperature (°C). To account for the metabolic effect, a reduction factor is incorporated into the model to account for the reduction in metabolic activity when the crop ages. When nitrogen content is simulated in the model, this factor can be related to the N content (van Keulen & Seligman, 1987). In the current model, the total rate of maintenance respiration is assumed to be proportional to the fraction of accumulated green leaves. This procedure for calculating the effect of age on maintenance respiration was used in the SUCROS model (Spitters et al., 1989) and was based on studies in which measured crop growth and canopy CO_2 assimilation data were analyzed using a simple simulation model (Louwerse et al., 1990; C.J.T. Spitters, CABO, unpublished data).

3.6   Growth Respiration

The carbohydrates in excess of the maintenance costs are available for conversion into structural plant material. Following the reactions in the biochemical pathways of the synthesis of dry matter compounds (carbohydrates, lipids, proteins, organic acids, and lignin from glucose (CH_2O)), Penning de Vries et al. (1974) have derived the assimilate requirements for these compounds. From the composition of the dry matter, the assimilate requirements for the formation of new tissue can be calculated. Typical values for leaves, stems, roots, and storage organs have been presented by Penning de Vries et al. (1989). The average carbohydrate requirements for the whole crop is calculated by weighting the coefficients with the fraction of dry matter allocated to the respective organs.

3.7   Daily Dry Matter Production

The daily growth rate (G_p, kg dry matter ha^{-1} d^{-1}) is calculated as:

$$G_p = [A_d \cdot (30/44) - R_m + R_t] / Q \qquad (3.13)$$

where A_d is the daily rate of gross CO_2 assimilation (kg CO_2 ha^{-1} d^{-1}), R_m is the maintenance respiration costs (kg CH_2O ha^{-1} d^{-1}), R_t is the amount of available stem reserves for growth (kg CH_2O ha^{-1} d^{-1}), and Q is the assimilate requirement for dry matter production (kg CH_2O kg^{-1} dry matter).

3.8   Dry Matter Partitioning

The total daily produced dry matter is partitioned among the various groups of plant organs (leaves, stems, storage organs, and roots) according to partitioning coefficients (*pc*, kg dry matter organ per kg dry matter crop) defined as a function of the phenological development stage (*DVS*):

$$pc_k = f(DVS) \qquad (3.14)$$

where k represents a particular organ. The dry matter is first distributed between the shoot and root, after which the shoot fraction is divided between stems, leaves and storage organs. The growth rate of plant organ group k (G_k) is thus obtained by multiplying the total potential growth rate (G_p, Equation 3.13) by the fraction allocated to that organ group $k(pc_k)$:

$$G_k = G_p \cdot pc_k \qquad (3.15)$$

The partitioning functions used in the model were derived from data sets collected by Akita (unpublished) for IR64 (Fig. 3.7). These partitioning functions have been found to be remarkably constant in experiments with varieties of different durations (Kropff *et al.*, 1993). Total dry weights of the plant organs are obtained by integrating their daily growth rates over time. The weight of rough rice is calculated from the panicle dry weight, the grain fraction in the panicle (0.86–0.90) and the moisture content of the grains (14%).

The fraction of allocable stem reserves is calculated as the difference between the maximum stem weight (around the time of flowering) and the stem weight at physiological maturity, expressed as a fraction of the maximum stem weight.

3.9   Sink Limitation

In grain crops the carbohydrate production in the grain-filling period can be higher than the storage capacity of the grains, which is determined by the number of grains per unit area and the maximum growth rate of the grains. This may result in the accumulation of assimilates in the leaves causing

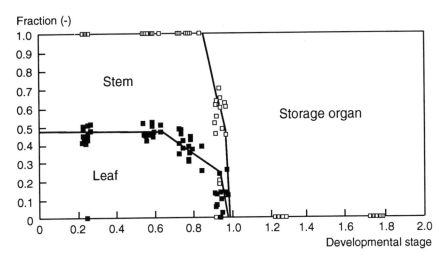

Fig. 3.7. The dynamic dry matter partitioning functions derived for IR64 (data from Akita, unpublished).

reduced rates of CO_2 assimilation through a feedback mechanism (Barnett & Pearce, 1983). This can be very important in rice when it is grown in extreme environments, as both low and high temperatures before flowering can induce spikelet sterility which results in a low sink capacity (Yoshida, 1981).

In wheat it has been found that the size of the spike at flowering is proportional to the number of grains that are formed (Fischer, 1985), and that spike size is closely correlated with the amount of growth of the crop during the spike formation period. The amount of growth over this period depends both on the duration of the period, which is influenced by temperature, and the crop growth rate, which is mainly influenced by solar radiation, and to a lesser extent, temperature. Similar relationships have been found in rice (Yoshida & Parao, 1976). Islam & Morison (1992) used a similar relationship in relating rice yields in Bangladesh to the 'photothermal quotient' (Q), the ratio of solar radiation in the 30 days prior to flowering to the mean temperature over the same period minus a base temperature.

In experiments at IRRI, we have found a good relationship between the total crop growth over the period from panicle initiation to 50% flowering and the number of spikelets at flowering (Fig. 3.8). This relationship holds across the wet and dry seasons, for levels of nitrogen application ranging from 0 to 285 kg ha^{-1}, from planting densities ranging from 25 to 125 plants m^{-2}, and for severe drought stress. A similar relationship is also found at the tiller level, so that the number of spikelets per tiller can be explained by the growth of each tiller during the period in which the panicle for that tiller is formed. The effects

Spikelets m⁻²

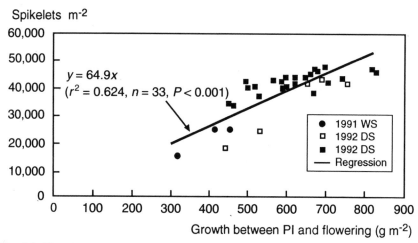

Fig. 3.8. The relationship between spikelet numbers and crop growth between PI and flowering (data from Cassman, Kropff, Torres and Liboon (symbols ●, ■) and S. Peng (□, IRRI), cultivar IR72).

of solar radiation, temperature, nitrogen, competition, and water, on spikelet formation, therefore, seem to be able to be integrated by their effects on crop growth over the panicle formation period. We have called the slope of this relationship the 'spikelet formation factor' (g). For a given variety, the relationship is remarkably consistent, although there do appear to be differences between varieties. For IR72, for example, g has a value of about 65 spikelets g⁻¹ total dry matter, but ranges between about 45 and 70 spikelets g⁻¹ in a number of varieties used in experiments at IRRI.

In the model, the amount of growth from panicle initiation (defined as DVS = 0.65) to first flowering (defined as DVS = 1.0) is tracked, and the number of spikelets at flowering (S_f) is calculated as the product of this growth and g, i.e.

$$S_f = \sum_{i=P}^{F} G_i \cdot g \qquad (3.16)$$

where P and F are the dates of panicle initiation and first flowering, respectively, and G_i is increment in shoot weight on day i. Thus, at flowering, a certain number of spikelets have been produced, which determines the maximum yield (Y_{max}, kg ha⁻¹) that can be achieved:

$$Y_{max} = S_f \cdot f_g \cdot G_{max} \cdot 10^4 \qquad (3.17)$$

where S_f is the number of spikelets at flowering (spikelets m⁻²), f_g is the fraction of these that form viable grains, and G_{max} is the grain size (mg per grain), assumed to be constant for a variety (Yoshida, 1981). The actual grain yield (Y, kg ha^{-1}) depends on the amount of assimilate produced in the period from flowering to maturity (defined as $DVS = 2.0$) plus any translocated assimilate from stem reserves, provided Y does not exceed Y_{max}, i.e.

$$Y = \sum_{i=F}^{M} (G_i + T_i), \qquad Y \leq Y_{max} \tag{3.18}$$

where F and M are the dates of flowering and maturity, respectively, and G_i and T_i are the increment in shoot weight and the amount of assimilate translocated, respectively, on day i. The model terminates the simulation when either $Y = Y_{max}$ or when maturity is reached, whichever occurs first.

3.10   Leaf Area Development

As discussed in Section 3.5, 50% of the green stem area index (SAI) is added to the leaf area index (LAI), to account for the lower photosynthetic activity of the stem. The SAI is calculated from the stem dry weight using the specific green stem area ($SSGA$) which was estimated from experimental data. After flowering, $SSGA$ is reduced because of death of sheath tissue.

For a closed canopy, the LAI is calculated from the leaf dry weight using the specific leaf area (SLA, m^2 leaf kg^{-1} leaf). When the LAI is less than 1.0, the plants grow exponentially as a function of the temperature sum:

$$LAI_{ts} = N L_{p,0} \exp (R_1 ts) \tag{3.19}$$

where LAI_{ts} is the leaf area index (m^2 leaf m^{-2} ground) at a specific temperature sum (ts, degree days, °Cd) after emergence, N the number of plants m^{-2}, $L_{p,0}$ the initial leaf area per plant at seedling emergence (m^{-2} plant^{-1}), and R_1 the relative leaf area thermal growth rate ((°Cd)$^{-1}$). The temperature sum ts is calculated using a base temperature of 8°C. The exponential phase ends when the portion of assimilates allocated to non-leaf tissue sharply increases, or when mutual shading becomes substantial. As an approximation for this transition, we used an arbitrary value of $LAI = 1.0$ as the end of the exponential growth period when leaves start to overlap.

In the seedbed of transplanted rice, development of LAI is first simulated using Equation 3.19. If LAI becomes larger than 1.0, further development is simulated using the SLA concept. At the date of transplanting, the duration of the transplanting shock period, in which there is no LAI growth, is calculated from the seedling age (Fig. 3.9). After the transplanting shock period, growth becomes exponential again, until the LAI exceeds 1.0, after which the SLA concept is again used. In case of direct-seeded rice, the date of transplanting is set equal to the date of seeding. Based on experimental results at IRRI, we have assumed that there is no effect of CO_2 on *SLA* (Ingram, pers. comm.).

To account for leaf senescence, a relative leaf death rate (RLDR, g g^{-1}) is defined as a function of the development stage. In the model, the existing green leaf weight is multiplied by RLDR to calculate the weight loss of the leaves (*LLV*). The reduction in leaf area is calculated from the loss of leaf weight using the specific leaf area (*SLA*).

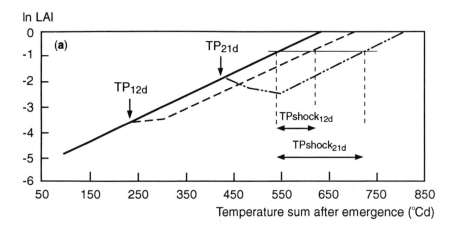

In LAI

Temperature sum after emergence (°Cd)

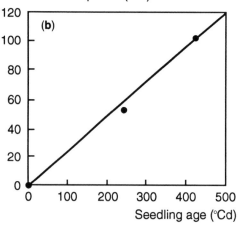

Transplanting shock for leaf area development (°Cd)

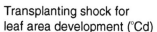

Seedling age (°Cd)

Fig. 3.9. (a) The relationship between the natural logarithm of leaf area of young rice plants and temperature sum (°Cd). Data include direct-seeded plants and seedlings transplanted at 12 days and 21 days after emergence. (From an unpublished experiment carried out in the 1991 wet season at IRRI, using the variety IR72, Torres, Liboon, Kropff & Cassman.) **(b)** Relation between the transplanting shock effect on leaf area development in rice expressed as a period where no growth occurs (TSHCKL,°Cd) and the seedling age at transplanting, also expressed in degree days. Data are from a wet season 1991 experiment with IR72 at IRRI, Los Baños, The Philippines (Torres, Liboon, Kropff & Cassman, IRRI, unpublished).

3.11   Input Requirements

Many of the parameters used in the model are species-specific and do not vary significantly between rice genotypes. The parameters for which we have found genetic variation are those related to phenological development, assimilate translocation, and spikelet development. A variety can therefore be defined by choosing appropriate values for these parameters. Environmental and crop management data that are required are:

IDOYS	Date of seeding
IDOYTR	Date of transplanting (equals date of seeding for direct-seeded rice)
NPLSB	Approximate number of plants per m^2 in seedbed
NH	Number of hills per m^2
NPLH	Number of plants per hill
RDTT	Daily solar radiation as function of the day of year
TMAXT	Maximum daily temperature as function of the day of year, and
TMINT	Minimum daily temperature as function of the day of year.

For simulation of yield potential, the initial relative growth rate of leaf area (RGRL) was set at 0.009 $(°Cd)^{-1}$. A reference time course of leaf nitrogen content in the leaves must also be input; to simulate potential yield, we used the leaf N content of the variety IR72 in an optimally managed experiment (225 kg ha^{-1} N) in the 1992 dry season at IRRI.

3.12   Carbon Balance Check

The model contains a carbon balance check, to ensure that total net assimilated carbon exactly equals the carbon fixed in dry matter and the carbon lost as a result of growth and maintenance respiration. The model gives an error message if the amount of carbon not accounted for is more than 0.1% of the total assimilated carbon.

3.13   Run Control

The model is available in both Turbo Pascal (Borland, v7.0) and FORTRAN (Microsoft, v5.1) versions, the former of which is able to run using the user-friendly shell GUMAYA (Matthews, Hunt & van den Berg, unpublished). The DSSAT3 data format is used when the model is operated through the shell. Further details of GUMAYA are given in Chapter 7.

3.14   Model Evaluation

The model was calibrated using data from experiments carried out at IRRI in
the 1991 wet season (1991 WS) and the 1992 dry season (1992 DS), and
evaluated using data from previous experiments at IRRI (Akita, unpublished

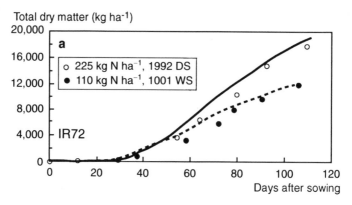

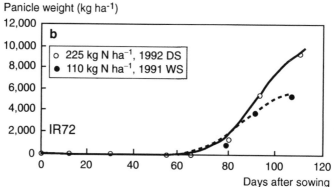

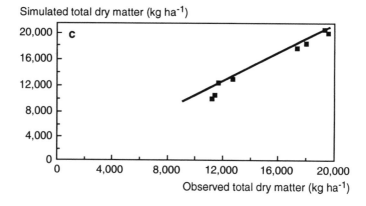

data), and also from experiments in different countries collaborating on the SIC2 project, details of which are given in subsequent chapters.

In the IRRI experiments, two improved semi-dwarf *indica* rice genotypes (IR72 and IR58109-113-3-3-2) were grown at three N levels during the 1991 wet season and 1992 dry season at the IRRI Research Farm in Los Baños, The Philippines. Nitrogen treatments were as follows: in the WS, 0, 80, and 110 kg N ha^{-1}; in the DS, 0, 180, and 225 kg N ha^{-1}. The rates in the highest N treatments were 30 kg N ha^{-1} (WS) and 105 kg N ha^{-1} (DS) greater than the current recommendations at IRRI, and included a late application at flowering to maintain leaf N status during grain filling. Seedlings were transplanted on 13 July (WS) and 16 January (DS). Treatments were arranged in a split plot design with N rates as the main plots, varieties as subplots, and four replicates. At regular

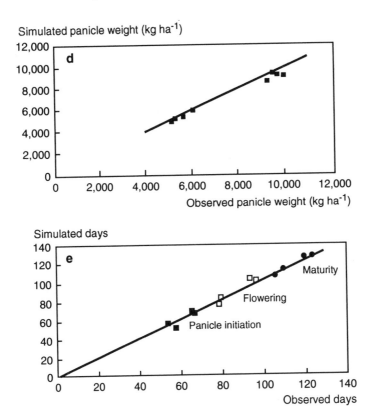

Fig. 3.10. (and opposite) Comparison of simulated (lines) and observed (symbols) values of various variables in wet and dry season experiments at IRRI. (a) Seasonal total crop dry weight for IR72; (b) seasonal panicle weights for IR72; (c) final harvest total crop dry matter for four varieties (IR72, IR58109, IRLB9104 and IRLB9205); (d) final harvest panicle weights for the four varieties; (e) phenological stages for the four varieties. Experimental data from Cassman, Kropff, Peng, Torres and Liboon, IRRI (unpublished).

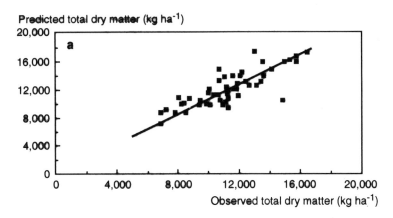

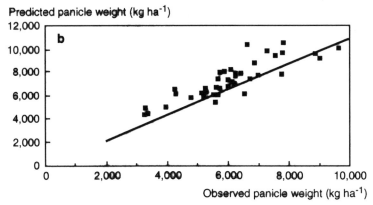

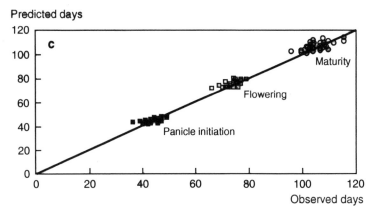

Fig. 3.11. (and opposite) Simulated and observed total dry weight, panicle dry weight and phenological events for IR58 (a,b,c) and IR64 (d,e,f). Data from several experiments conducted by Akita (unpublished data, IRRI).

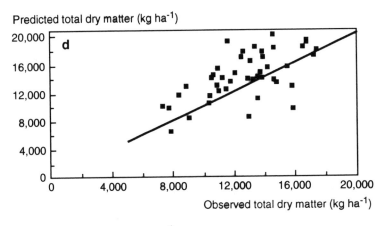

Predicted total dry matter (kg ha⁻¹)

Observed total dry matter (kg ha⁻¹)

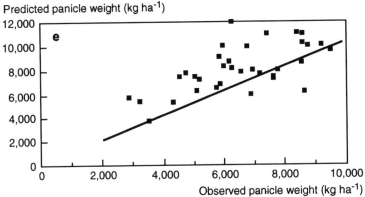

Predicted panicle weight (kg ha⁻¹)

Observed panicle weight (kg ha⁻¹)

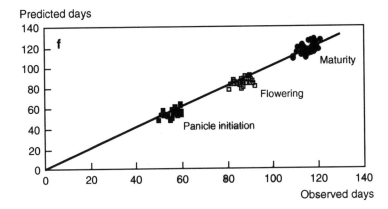

Predicted days

Observed days

intervals, 12-hill samples were taken from each plot to monitor *LAI*, dry matter, and N content in the leaf blades, stems, and panicles.

The model was calibrated using the data from all high N treatments of these experiments. The model was able to describe seasonal growth, final dry matter production and yield, and phenological events accurately for both seasons (Fig. 3.10 a–e).

Subsequently, the model was evaluated using a wide range of experimental data for IR58 and IR64 from several years and a wide range of treatments. Relative yield differences between experiments were adequately simulated (Fig. 3.11), although for IR64 yields were consistently overestimated. This could have been the result of the lower proportion of filled grains in the experiments (60%), whereas the model assumes a potential of 90%.

3.15   Model Responses to Temperature and CO_2

The model was evaluated with respect to its response to temperature and CO_2 using data from the closed outdoor chamber experiments of Baker *et al.* (1990c). As the actual experimental conditions are unknown, the response of relative yields (relative to the overall experimental mean) to mean temperature were compared. Figure 3.12 shows that the model predicts the response of grain yield to temperature and CO_2 reasonably accurately up to mean daily temperatures of 32°C, but overestimates yields in the range 32–36°C. This could be due to errors in the estimation of spikelet sterility at these temperatures; the model uses a relation based on a *japonica* variety from Japan, but it is known that considerable genotypic variation exists for this characteristic (Satake & Yoshida, 1978). The variety used by Baker and his colleagues may be more sensitive to temperatures in this range than the variety used in the model.

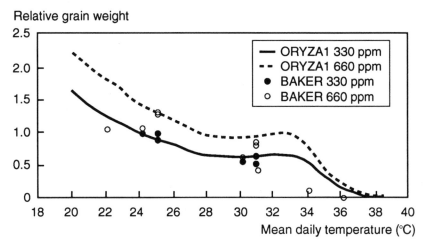

Fig. 3.12. Observed (symbols; Baker *et al.*, 1990c) and simulated (lines) effects of temperature and CO_2 on rice grain yield, relative to current conditions (24°C/330 ppm).

The Rice Crop Simulation Model SIMRIW and Its Testing   **4**

T. Horie[1], H. Nakagawa[1], H.G.S. Centeno[2], and M.J. Kropff[2]

[1]*Faculty of Agriculture, Kyoto University, Kyoto 606, Japan;* [2]*International Rice Research Institute, P.O. Box 933, 1099 Manila, The Philippines*

4.1   Introduction

SIMRIW (SImulation Model for RIce–Weather relations) is a simplified process model for simulating growth and yield of irrigated rice in relation to weather. The model was developed by a rational simplification of the underlying physiological and physical processes of the growth of the rice crop (Horie, 1987). Because of this, it requires only a limited number of crop parameters which can be obtained easily from well defined field experiments, and, hence, is applicable to a wide range of environments. Although SIMRIW predicts the potential yield that can be expected from a given cultivar under a given climate, actual farmers' yield at a given location or district can also be obtained by multiplying the potential yield by a 'technological coefficient' that characterizes the current level of rice cultivation technology (i.e. fertilizer applications, soil, water, weeds and pest management, etc.) at the location. It has been shown that the model can satisfactorily explain the locational variability of rice yields in the USA and Japan based on the respective climates (Horie, 1987), and that it also explains the yearly variations in yield at various districts in Japan based on the weather (Horie *et al.*, 1992). SIMRIW, in combination with Meash Weather Information Systems is being used for growth and yield forecasting of regional rice in some of the prefectures in Japan (Horie *et al.*, 1992).

SIMRIW was expanded to allow for its application for impact assessments of global environmental change on rice growth and yield in various regions, by incorporating into it the processes related to effects of changes in CO_2 concentration in the atmosphere and high temperature on rice (Horie, 1993). This chapter describes the structure of SIMRIW and results of its validation.

4.2   General Overview of the Model

SIMRIW is a simplified process model for simulating the potential growth and yield of irrigated rice in relation to temperature, solar radiation, and CO_2 concentration in the atmosphere. The model is based on the principle that the grain yield (Y_G, g m^{-2}) forms a specific proportion of the total dry matter production (W_t, g m^{-2}) of a crop:

$$Y_G = hW_t \tag{4.1}$$

in which h is the harvest index.

It has been shown that crop dry matter production is proportional to the amount of quanta of photosynthetically active radiation (PAR) absorbed by a crop canopy (Shibles & Weber, 1966). A strong relationship also exists between dry matter production and the amount of short-wave radiation intercepted (Monteith, 1977; Gallagher & Biscoe, 1978). Horie & Sakuratani (1985) showed that this is also true in rice and that the proportionality coefficient, the conversion efficiency from solar radiation to biomass, is constant until the middle of the ripening stage and thereafter decreases curvilinearly. Moreover, they concluded from both simulations and experiments that the conversion efficiency is practically unaffected by climatic conditions in a wide range of environments.

This relationship is described in SIMRIW as follows:

$$dW_t/dt = C_s I_s \tag{4.2}$$

where C_s is the conversion efficiency of absorbed short-wave radiation to rice crop biomass (g dry matter MJ^{-1}), I_s is the absorbed radiation per unit time (MJ m^{-2} d^{-1}), and t is the time unit (d). Since the time coefficient (the reciprocal of the relative growth rate) of growth after transplanting into the field is more than five days for rice, it is sufficiently accurate to integrate Equation 4.2 with a time interval of one day. Hence equation 4.2 can also be represented as:

$$\Delta W_t = C_s S_s \tag{4.3}$$

in which ΔW_t is the daily increment of the crop dry weight (g m^{-2} d^{-1}), and S_s is the daily total absorbed radiation (MJ m^{-2} d^{-1}).

Figure 4.1 shows schematically how the processes of growth, development and yield formation in rice are modeled in SIMRIW. The x-axis of Fig. 4.1 represents the development index (DVI) which is a measure of the crop development stage on a given day. The quantities h, C_s and S_s are functions of the environment, crop developmental stage, and growth attributes, respectively. These functions and parameters have been determined from field experiment data for eight major cultivars of both *japonica* and *indica* rice varieties grown under widely different environmental conditions. Only the principal functions of the model are described in the subsequent sections of this report; fuller details of the derivation of each functional relationship and parameters of the model are given in Horie (1987).

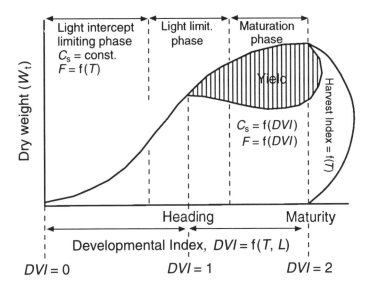

Fig. 4.1. Schematic representation of processes of growth, development and yield formation of rice used in SIMRIW (Horie *et al.*, 1992). (*T* is temperature, *L* is photoperiod (h), *F* is leaf area index, C_s is conversion efficiency.)

4.3 Phenological Development of the Crop

The developmental processes of the rice crop, such as ear initiation, booting, heading, flowering, and maturation are strongly influenced both by the environment and by the crop genotype. In SIMRIW, these are described by the developmental index (*DVI*) in a similar way to that employed by de Wit *et al.* (1970). This variable is defined as 0.0 at crop emergence, 1.0 at heading, and 2.0 at maturity. Thus, the development stage at any time in the life of the crop is represented by a value between 0.0 and 2.0.

The value of *DVI* is calculated by summing the developmental rate (*DVR*) with respect to time:

$$DVI_t = \sum_{i=0}^{i=t} DVR_i \qquad (4.4)$$

where DVI_t is the developmental index at day t, and DVR_i is the developmental rate on the i-th day from emergence. Daylength and temperature are known to be the major environmental factors determining *DVR*. In rice, *DVR* from emergence to heading can be represented as a function of daylength (*L*) and daily mean temperature (*T*) as follows (Horie & Nakagawa, 1990):

$$
\begin{aligned}
\text{for} \quad & DVI \leq DVI^* & & DVR = 1 / \langle G_v \{1.0 + \exp[-A(T - T_h)]\}\rangle \\
& DVI > DVI^* \text{ and } L \leq L_c & & DVR = \{1 - \exp[B(L - L_c)]\} / \\
& & & \qquad G_v\{1 + \exp[-A(T - T_h)]\} \\
& DVI > DVI^* \text{ and } L > L_c & & DVR = 0
\end{aligned}
\qquad (4.5)
$$

where DVI^* is the value of *DVI* at which the crop becomes sensitive to photoperiod, L_c is the critical daylength (h), T_h is the temperature at which *DVR* is half the maximum rate at the optimum temperature (°C), G_v is the minimum number of days required for heading of a cultivar, under optimum conditions of temperature and photoperiod, and where nutrients, pests and diseases are not limiting, and A, B are empirical constants.

The values of the constants in Equation 4.5 differ among cultivars. In general, late-maturing cultivars have a larger G_v, while cultivars with a higher sensitivity to photoperiod have a larger B and smaller L_c. These parameters were determined for eight varieties by applying the 'simplex' method, a trial-and-error method to estimate parameters of non-linear functions (Horie & Nakagawa, 1990), for field data on rice phenology obtained in various regions in Japan. Figure 4.2 shows the *DVR* of the variety IR36 in the photoperiod sensitive phase as a function of *T* and *L*. When refitted to the datasets used in its calibration, the model could describe the days to heading of field-grown rice with a standard error of only 3–4 d over a range of 75 to 145 d (Fig. 4.3).

On the basis of observation that the period from emergence to heading was shortened by 4% in rice under doubled CO_2 conditions (Nakagawa *et al.*, 1993), the basic vegetative period (G_v) of Equation 4.5 was given by the following equation as a function of CO_2 concentration:

$$G_v = G[1 - 0.000114(C_a - 350.0)] \qquad (4.6)$$

where C_a is atmospheric CO_2 concentration (ppm), and G is the value of G_v at $C_a = 350$ ppm.

DVR (d^{-1})

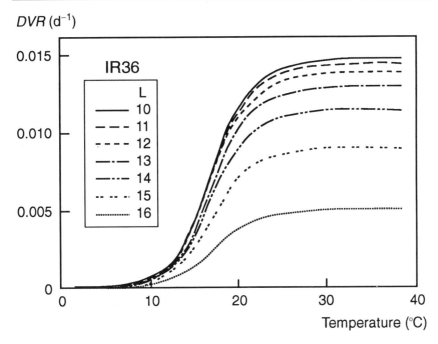

Fig. 4.2. Development rate (DVR) of the variety IR36 as a function of daily mean temperature (T, °C) and photoperiod (L, h), after calibration of the relevant parameters of Equation 4.5 from field data (Horie, 1987).

The following equation is used to describe the rate of development from heading to maturity ($1 < DVI < 2$):

$$DVR = \{1 - \exp[-K_r(T - T_{cr})]\} / G_r \qquad (4.7)$$

where G_r is the minimum number of days for grain-filling period, under optimum conditions of temperature and photoperiod, and where nutrients, pests, and diseases are not limiting, and K_r and T_{cr} are empirical constants.

4.4 Dry Matter Production

The amount of radiation absorbed by the canopy (S_s) is a function of leaf area index (F), and the structure and optical properties of the canopy. In SIMRIW, S_s is calculated using the equation of Monsi & Saeki (1953):

$$S_s = S_0 \{1 - r - (1 - r_0) \exp[-(1 - m) k F]\} \qquad (4.8)$$

where S_0 is the daily incident solar radiation (MJ m^{-2} d^{-1}), r and r_0 are the

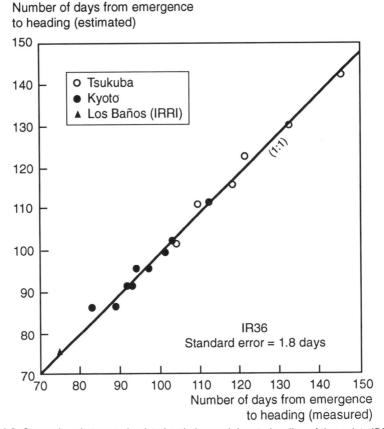

Number of days from emergence
to heading (estimated)

o Tsukuba
● Kyoto
▲ Los Baños (IRRI)

(1:1)

IR36
Standard error = 1.8 days

Number of days from emergence
to heading (measured)

Fig. 4.3. Comparison between simulated and observed days to heading of the variety IR36 under different environmental conditions (Horie, 1987).

reflectance of canopy and the bare soil, respectively, m is the scattering coefficient, and k is the extinction coefficient of the canopy to daily short-wave radiation. The canopy reflectance (r) is given by the following equation (Research Group of Evapotranspiration, 1967):

$$r = r_f - (r_f - r_0) \exp(-F/2) \tag{4.9}$$

in which r_f is the reflectance when the surface is completely covered by the vegetation. From Horie & Sakuratani (1985), the following values were adopted for the parameters of the above functions: $k = 0.6$, $m = 0.25$, $r_f = 0.22$ and $r_0 = 0.1$.

Daily dry matter production is calculated by multiplying the S_s value by an appropriate value of the radiation conversion efficiency C_s (Equation 4.3). As has been shown in Horie & Sakuratani (1985) C_s is constant until the middle of the grain-filling stage, after which it decreases gradually toward zero at maturity. This pattern is simulated using the following equations:

for $\quad 0.0 < DVI < 1.0 \quad C_0 = C$
$\quad\quad 1.0 \leq DVI < 2.0 \quad C_0 = C(1 + B) / \{1 + B \exp [(DVI - 1) / t]\}$ (4.10)

in which C_0 is the radiation conversion efficiency at 330 ppm CO_2 (g MJ^{-1}), and C, B, t are empirical constants. However, C_s is also influenced by the CO_2 concentration of the atmosphere. Horie (1993) showed that the following equation fits the CO_2 response of the radiation conversion efficiency in rice:

$$C_s = C_0 \{1 + R_m(C_a - 330) / [(C_a - 330) + K_c]\} \quad\quad (4.11)$$

where R_m is the asymptotic limit of relative response to CO_2, and K_c is an empirical constant (ppm).

In SIMRIW, the expansion of leaf area is modeled independently of leaf weight, for reasons outlined in Horie et al. (1979). It is well documented that CO_2 enrichment has little or no effect on leaf area development in rice (Imai et al., 1985; Baker et al., 1990a; Nakagawa et al., 1993). Under optimal cultivation, it is reasonable to assume that water and nutrients are not limiting factors in the expansion of leaf area, and that the main governing factor is temperature. In SIMRIW, the relationship between relative growth rate of leaf area index (F) and daily mean temperature (T) for the period before heading is given as:

$$1/F \times dF/dt = A\{1 - \exp [-K_f (T - T_{cf})]\} [1 - (F/F_{as})^h] \quad\quad (4.12)$$

in which A is the maximum relative growth rate of LAI (m^2 m^{-2}), obtained under optimum conditions where temperature, solar radiation, nutrients, pests, and diseases are not limiting, T_{cf} is the minimum temperature for LAI growth (°C), F_{as} is an asymptotic value of the leaf area index when temperature is non-limiting (m^2 m^{-2}), and K_f and h are empirical constants.

In rice, it is commonly observed that LAI attains a maximum at about the heading stage and thereafter declines during the grain filling stage due to cessation of new leaf growth coupled with leaf senescence. However, the physiology of leaf senescence is not well understood. For this reason, changes in leaf area index from the time just before heading to crop maturity is described in SIMRIW by an empirical function of the crop developmental index (DVI).

4.5   Yield Formation

Figure 4.4 shows that the relationship between the dry weight of the brown rice (unprocessed grain), W_y, and that of the whole crop (W_t, including roots) is

Dry weight of brown rice (t ha⁻¹)

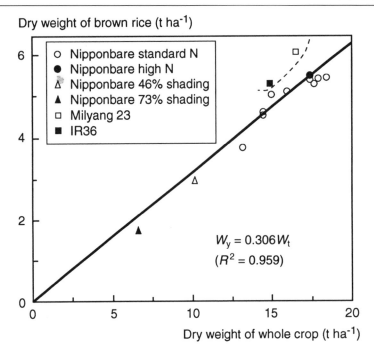

Fig. 4.4. Relationship between the dry weight of brown rice (W_y) and that of whole crop (W_t) for the variety Nipponbare, grown under different environmental conditions (Horie, 1988). Data on two other high high-yielding varieties are also shown.

linear over a wide range of W_t, indicating that the harvest index (h) is constant. However, this is not always the case. The harvest index decreases if the fraction of sterile spikelets increases, or if the crop growth stops before completing its development due to cool summer temperatures or frost. The increase in the number of sterile grains brought about by cool temperatures at the booting and flowering stages is called cool-summer damage due to floral impotency, and the premature cessation of growth due to low temperatures in summer is called cool-summer damage due to delayed growth.

In SIMRIW, the harvest index h is represented as a function of the fraction of sterile spikelets (γ) and the crop developmental index (DVI) in order to take into account both types of cool-summer damage, as follows:

$$h = h_m(1 - \gamma)\{1 - \exp[-K_h(DVI - 1.22)]\} \quad (4.13)$$

where h_m is the maximum harvest index of a given cultivar, obtained under optimum conditions where temperature, solar radiation, nutrients, pests and diseases are not limiting, and K_h is an empirical constant. Equation 4.13 implies that the harvest index decreases as γ increases due to cool temperature at the booting and flowering stages, or according to the date of cessation of growth before full maturation ($DVI = 2.0$) due to cool summer temperatures.

Using the 'cooling degree-day' concept (Uchijima, 1976), the relation between daily mean temperature (T_i, °C) and the percentage sterility may be approximated by the following equation (Fig. 4.5):

$$\gamma = \gamma_0 - K_q\, Q_t^a \qquad (4.14)$$

where γ_0, K_q and a are empirical constants, and Q_t is the cooling degree-days, given by:

$$Q_t = \Sigma\, (22 - T_i) \qquad (4.15)$$

The summation of Equation 4.15 is made for the period of highest sensitivity of the rice panicle to cool temperatures ($0.75 \leq DVI \leq 1.2$).

Rice spikelets are also sensitive to high temperature, particularly at anthesis. Damage to the pollen occurs when the temperature at flowering is above approximately 35°C (Satake & Yoshida, 1978; Matsui & Horie, 1992). Figure 4.6 represents the relationship between the fraction of fertile spikelets

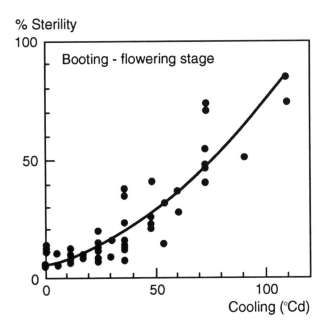

Fig. 4.5. Relation between cooling degree-days and percentage spikelet (γ) sterility of the variety Eiko between booting and flowering stages (Horie, 1988 constructed from data of Shibota *et al.*, 1990).

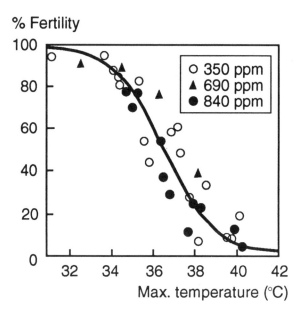

Fig. 4.6. Relation between average daily maximum temperature during the flowering period and spikelet fertility in the variety Akihikari acclimated to different CO_2 concentrations (Horie, 1993).

and the average daily maximum temperature over the flowering period ($0.96 \leq DVI \leq 1.22$) for Akihikari rice grown in a temperature gradient tunnel (Horie, 1993) with elevated and ambient CO_2 concentrations. Figure 4.6 indicates that CO_2 concentration has no effect on the temperature and fertility relationship. The relation shown in Fig. 4.6 may be approximated by

$$1 - \gamma = 1 - 1/\{1 + \exp[-0.853 (T_M - 36.6)]\} \qquad (4.16)$$

where T_M is the average daily maximum temperature during the flowering period. Daily maximum temperature is employed to account for rice spikelets usually flowering during the day time. Actual spikelet sterility is calculated as the minimum of Equations 4.14 and 4.16.

The model terminates when either *DVI* reaches 2.0 (full maturation) or when three consecutive days are encountered in which the mean temperature is below the critical temperature for growth. This critical temperature is taken as 12°C.

4.6   The SIMRIW Program

SIMRIW is programmed in FORTRAN, and consists of a main program to compute the equations described above and one subroutine to input weather data

(Fig. 4.7). SIMRIW also requires two external files: CROPARAM.DAT for specification of cultivar specific crop parameters, and a weather data file containing daily weather data. Results of the simulation are output to a file called RESULTS.SIM.

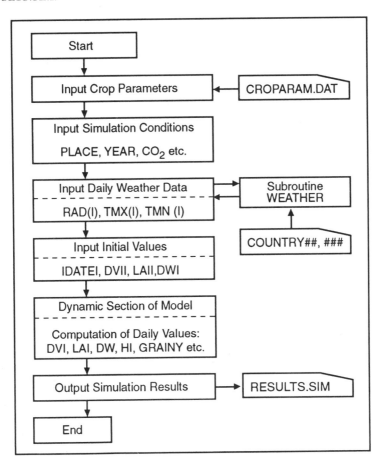

Fig. 4.7. Flow chart of the SIMRIW program.

In the current version of SIMRIW, 25 cultivar-specific parameters are required. Values of these parameters for eight rice cultivars are contained in the CROPARAM.DAT file; a menu allows users of the model to easily choose one of these cultivars for simulation. The most sensitive crop parameters were found to be those related to phenology, radiation conversion efficiency, and the initial values of crop developmental index (*DVII*) and leaf area index (*LAII*). Careful specification is needed for the values of *DVII* and *LAII*. General ranges of these values are suggested on the screen. SIMRIW requires daily values of maximum and minimum temperatures (°C) and daily solar radiation (MJ m^{-2} d^{-1}) in FSE-format (Fortran Simulation Environment; van Kraalingen *et al.*, 1991). For each run the name of the weather data file needs to be specified by the user.

4.7   Model Validation

The development of SIMRIW has been based on experimental data using *japonica* varieties from Japan, a temperate country, and not for *indica* varieties in tropical regions. The model has been shown to satisfactorily explain locational and yearly variations of rice yield in various regions in Japan (Horie, 1987, 1988, 1993; Horie *et al.*, 1992, and also Chapter 6 of this book). For this reason, the model validation in this chapter is focused on the ability of SIMRIW to predict growth and yield of rice under tropical conditions.

The same two data sets used in the testing of ORYZA1 (Chapter 3) were used to test SIMRIW. Experiment I was an optimum management experiment using the variety IR72, which was conducted at IRRI in the 1991 wet season and 1992 dry season (Kropff *et al.*, 1993), while Experiment II was a year-round time-of-planting experiment also at IRRI using the variety IR58 during 1987–1988 (Akita, unpublished). In Experiment I, the wet season crop was transplanted on 12 July 1991, and a total of 110 kg N ha^{-1} was applied in split applications, while the dry season crop was transplanted on 16 January 1992, and a total of 225 kg N ha^{-1} was applied, again in split applications. In Experiment II, seedlings of IR58 were transplanted at monthly intervals from March 1987 until March 1988, and standard crop management techniques employed. Crop parameters determined previously for IR36 were used to simulate the growth of IR72, because the two varieties have a similar phenology. Simulations of both experiments commenced at transplanting. Initial values, based on experimental measurements, used for Experiment I were *DVII*=0.12, *LAII*=0.03, and *DWI*=100 kg ha^{-1}, and for Experiment II, *DVII*=0.2, *LAII*=0.04 and *DWI*=100 kg ha^{-1}.

Figure 4.8 shows the simulated and observed growth curves of IR72 in Experiment I, along with the weather conditions. There was close agreement between simulated and observed curves of the whole crop and panicle, both in the wet and dry seasons. The observed days from transplanting to heading of IR72 were 67 and 68 days in the wet and dry seasons respectively, while the corresponding simulated values were 66 and 67 days, respectively. Thus, SIMRIW could predict almost perfectly the phenology of IR72.

Figure 4.9 shows the measured and simulated final panicle dry weight of IR58 as a function of transplanting date in Experiment II. Except for the December 1987 transplanting, the simulated yields were always higher than the measured yields. This is in contrast to the results in Experiment I (Fig. 4.8), which was optimally managed, including split applications of abundant nitrogen.

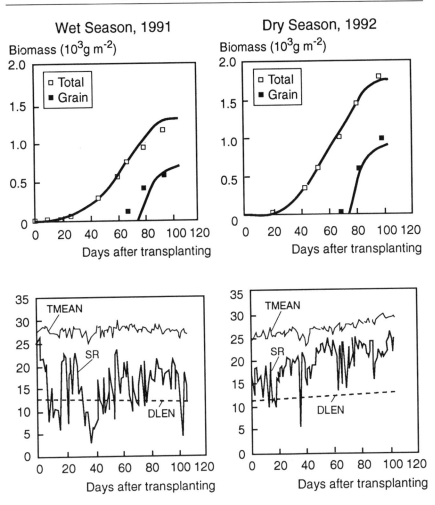

Fig. 4.8. Simulated and observed growth curves for the whole crop and the panicle of IR72 grown in the wet and dry seasons at IRRI, The Philippines. Daily mean temperature (TMEAN, °C), solar radiation (SR, MJ m⁻² d⁻¹) and daylength (DLEN, h) are also shown. (Unpublished data from Cassman & Kropff, IRRI.)

Since SIMRIW predicts the potential yield possible for a given cultivar under optimal conditions, the results shown in Figs 4.8 and 4.9 suggest that cultivation conditions for Experiment I were close to optimal, but were suboptimal in Experiment II.

Although the simulated yields in Experiment II were higher than the measured yields, the results shown in Fig. 4.9 indicate that SIMRIW can predict the

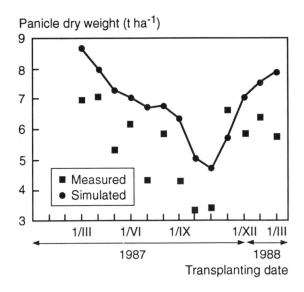

Fig. 4.9. Simulated and measured seasonal changes of panicle dry weight of IR58 transplanted at monthly intervals from March 1987 to March 1988 at IRRI, The Philippines. (Data from Akita, IRRI.)

seasonal variation of rice yields observed at IRRI. Indeed, a close linear relation between simulated and observed yields was obtained ($r^2 = 0.828$), with the exception of the December 1987 transplanting (Fig. 4.10). It seems, therefore, that SIMRIW, which was developed to predict the growth of temperate rice in relation to weather, can also satisfactorily simulate relative variation in the growth and yield of tropical rice. For this experiment, the predicted potential yield (Y_p) can be converted to actual yield (Y_a) by

$$Y_a = K Y_p \qquad (4.17)$$

in which K is a 'technology coefficient' representing the level of technology applied to the experiment. In applying the model to evaluate the effects of climate change, we have assumed the value of K is constant, and considered only the *relative* predicted changes in potential yield. It is recognized that K itself is likely to change as a function of time (see Chapter 8), so that by the middle of the next century it may well have a different value to that at present. However, it is outside the scope of this study to predict how improvements in management might affect future actual rice production; we have restricted the analysis, therefore, to a consideration of proportional changes in potential production.

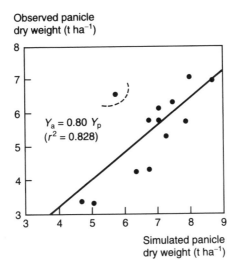

Fig. 4.10. Relation between simulated (Y_p) and measured (Y_a) panicle dry weight of IR58 transplanted at monthly intervals from March 1987 to March 1988 at IRRI, The Philippines. (Data from Akita, IRRI.)

4.8   Sensitivity Analysis of the Model to Environments

A sensitivity analysis of SIMRIW to conditions in the aerial environment was made by examining responses of simulated yield to daily mean temperature, solar radiation, and CO_2 concentration, under constant conditions of those environments over the entire growth season (Fig. 4.11). In this analysis, the cultivar Nipponbare was used, and the mean daily temperature was varied, starting at 18°C and increasing at intervals of 4°C up to 38°C. The diurnal range of temperature was set at 8°C. Under the same conditions each day, the optimum mean temperature for yield was found by simulation to be 22–23°C, below which the yield decreased sharply due to the increase of sterile spikelets from cool temperature damage. As the temperature increased above the optimum, yield declined more or less linearly up to about 30°C, from which it declined sharply. The initial linear decline of yield with temperature reflects the fact that, as the temperature increases, the total crop duration is shortened due to acceleration of phenological development. The sharp decline of yield above about 30°C is due to spikelet sterility from high temperature damage.

The overall response pattern of simulated yields to temperature is similar to the results of Munakata (1976) in which the effect of temperature on yields was

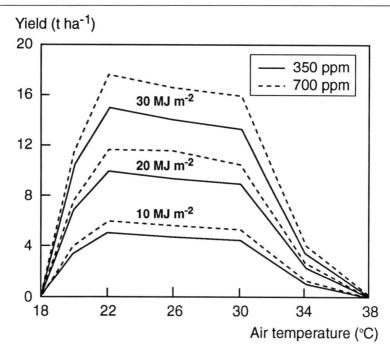

Fig. 4.11. Simulated yield responses of the variety Nipponbare to daily mean temperature, solar radiation, and CO_2 concentration, under constant environmental conditions. Daylength was set at 12 h in each case.

statistically analyzed by using data from long-term field experiments from various regions in Japan. The linear decline of simulated yield over the temperature range 22–30°C is similar to the results of Yoshida & Parao (1976) obtained from experimental data at IRRI.

As is expected from the use of the radiation conversion efficiency concept, simulated rice yields were proportional to solar radiation over the entire temperature range, which agrees with the experimental results presented by Yoshida & Parao (1976). The model predicts that doubling the level of CO_2 in the atmosphere increases rice yield by 25% over all temperature and solar radiation conditions.

Climate Change Scenarios   5

D. Bachelet[1], G.A. King[1], and J. Chaney[2]

[1]ManTech Environmental Technology, Inc., US EPA Environmental Research Laboratory, 200 SW 35th Street, Corvallis, Oregon 97333, USA; [2]Department of Botany and Plant Pathology, Oregon State University, Corvallis, Oregon, USA

5.1   Scenarios of Future Climate

A climate change scenario is defined as a physically consistent set of changes in meteorological variables, based on generally accepted projections of CO_2 and levels of other trace gases (Fischer *et al.*, 1994). Several approaches have been suggested for generating scenarios of future climate change caused by increases in greenhouse gas concentrations in the atmosphere (e.g. Robinson & Finkelstein, 1989; King, 1993).

Firstly, arbitrary scenarios (e.g. +1°C, +2°C, +4°C, and ±10% precipitation) have been used to determine the sensitivity of the system to climatic perturbations. These are essentially a basic sensitivity analysis testing of the crop simulation models as traditionally used in the past. These 'scenarios' are, of course, not tied to increases in radiative forcing in any physically detailed manner.

Secondly, historical analogs have also been used to determine the sensitivity of ecosystems to climate change (e.g. Smith & Tirpak, 1989; Rosenberg & Crosson, 1991). The limitation of using an analog period is that future change is likely to be faster and greater than any climatic change that has occurred over an extended period of time (more than a few months) in the instrumental record. Also, the causes of earlier climate fluctuations are different from the greenhouse-gas-induced forcing that will drive future climatic change. For instance, the warmer summers and cooler winters that existed in the northern hemisphere 9000 years ago were in large part likely caused by changes in solar insolation. Summer solar insolation was 8% greater than today while winter insolation was 8% less (Kutzbach, 1987). Greenhouse-gas-induced warming will likely result in warming in both summer and winter (Houghton *et al.*, 1990, 1992) although the issue is still under debate. Greater attention is now paid to daytime versus night-time temperatures because of their links to the changes in cloudiness, humidity, atmospheric circulation patterns, wind and soil moisture, which are likely to be significant over land areas (Houghton *et al.*, 1992; Kumar *et al.*, 1994).

Thirdly, physically-based climate models can be used to generate estimates of future climate. The most comprehensive models available are the General Circulation Models (GCMs), which simulate the circulation and physical properties of the atmosphere and surface meteorological variables over a three-dimensional spatial grid representing the earth–climate system. GCMs are based on the fundamental laws of atmospheric physics, although many of the atmospheric processes that take place at a scale smaller than the spatial resolution of the models are represented by simplified terms and equations (Houghton *et al.*, 1990, 1992). GCMs mechanistically simulate the climatic effects of increases in greenhouse gases in a physically consistent manner for the entire globe.

In this study, we have used the first and third approaches to evaluate the effects of changes in the climate on rice production. For the first approach, simulations were made for various levels of CO_2, since CO_2 is the most certain and virtually uniform change predicted. The effects of fixed increments of temperature were then simulated, along with combinations of CO_2 and temperature corresponding to moderate CO_2 and temperature effects, using IPCC scenarios as a guideline (Houghton *et al.*, 1990, 1992). Finally, as they are the least certain, the effect of the changes in temperature and solar radiation predicted by three GCMs were simulated. In this way, the simulations are stratified according to the certainty of the environmental change being considered.

5.2 GCM Predictions

5.2.1 General limitations to the reliability of GCM predictions

GCMs have been used extensively to provide potential climatic change scenarios (e.g. Grotch, 1988; Gutowski *et al.*, 1988; Parry *et al.*, 1988; Smith & Tirpak, 1989; Cohen, 1990) even though they have significant inherent limitations in projecting regional climate patterns (e.g. Houghton *et al.*, 1992; Bachelet *et al.*, 1993; King, 1993). The most significant limitations include: (i) poor spatial resolution; (ii) inadequate coupling of atmospheric and oceanic processes; (iii) poor simulation of cloud processes; and (iv) inadequate representation of the biosphere and its feedbacks. One of the difficulties in using GCMs is that there are many different models being used to simulate climate change and several different runs are now available from several modeling groups as they revise their GCMs. We chose the 'standard runs' that have been used most frequently in the literature (e.g. Parry *et al.*, 1988; Leemans & Solomon, 1993; Rosenzweig & Parry, 1994) for the sake of comparison, even though a few runs at higher spatial resolution are now available.

Another limitation of the GCMs is that most available runs are equilibrium rather than transient (time-dependent simulations). The rate of climate change is of critical importance for mitigation efforts and at least four coupled atmosphere–ocean GCMs have now been completed (Houghton *et al.*, 1992). Again, since we wanted our results to be compatible with other projects, we decided to only use equilibrium runs which were the only ones available when the project started five years ago.

In general, GCMs can at best be used to suggest the likely direction and rate of change, because they still have significant problems in simulating current climate (Houghton *et al.*, 1992).

5.2.2   Current climate predictions

In a previous study (Bachelet *et al.*, 1993), three GCM models were compared with each other and also to long-term means from climatological records for all the meteorological stations (US airfield summaries, Hatch, 1986) present in each GCM grid cell across Asia. Details of these three GCMs – the Geophysical Fluid Dynamics Laboratory model (GFDL, Wetherald & Manabe, 1988), the Goddard Institute for Space Studies model (GISS, Hansen *et al.*, 1988), and the United Kingdom Meteorological Office model (UKMO, Wilson & Mitchell, 1987) – are shown in Table 5.1. The baseline runs that were used in the study assumed current atmospheric CO_2 levels ($1 \times CO_2$), although these differed slightly between models (Table 5.1). Model output in the form of monthly

Table 5.1.   Major features of the three GCMs used in this study.

	GCM		
	GFDL	GISS	UKMO
Source laboratory	Geophysical Fluid Dynamics Laboratory	Goddard Institute for Space Studies	United Kingdom Meteorological Office
Reference	Wetherald & Manabe (1988)	Hansen *et al.* (1988)	Wilson & Mitchell (1987)
Horizontal resolution (lat. × long.)	$4.44° \times 7.5°$	$7.83° \times 10.0°$	$5.0° \times 7.5°$
Vertical resolution (no. of layers)	9	9	11
Diurnal cycle	No	Yes	Yes
Base $1 \times CO_2$ (ppm)	300	300	323
Change in global temperature (°C)	+4.0	+4.2	+5.2
Change in global precipitation (%)	8	11	15

statistics corresponded to an average of ten years of predictions for the GFDL and GISS models and 15 years for the UKMO.

While predictions of current temperature were relatively accurate ($r^2 = 0.8$–0.9), for monthly rainfall there was very poor agreement, not only between model predictions and measured records, but also between the model predictions themselves (Table 5.2). A detailed analysis of this study can be found in Bachelet *et al.* (1992, 1993). The lack of agreement between model output and observed data is due to several causes which are discussed below.

Table 5.2.   Results from a regression analysis comparing GCM predictions of monthly average temperature and monthly rainfall to long-term averaged climatic records for all the meteorological stations (airfield summaries) included in each of the GCM grid cells over Asia. The average number of meteorological stations included in each GCM cell is also shown. (Adapted from Bachelet *et al.*, 1993)

	GCM		
	GFDL	GISS	UKMO
No. of stations	5	10	6
Monthly average temperature (°C)			
Slope	1.29	0.97	1.24
y-intercept	−10.71	−0.33	−7.02
r^2	0.86	0.90	0.83
Monthly rainfall (mm)			
Slope	0.22	0.17	0.42
y-intercept	98.08	122.74	82.00
r^2	0.15	0.06	0.12

5.2.3   GCM limitations for current climate

A number of problems arise when general circulation model output and data are compared over a region. First, models generally produce results at grid points that, due to computational costs, are typically separated by several hundreds of kilometers. In such cases it is not immediately obvious what the temperature or rainfall at a gridpoint really represents. Because of this relatively crude areal resolution, only a few gridpoints (80 to 247 for Asia) were used to estimate averages for subcontinental intercomparisons (Joyce *et al.*, 1990) and were compared to data from three to ten times as many meteorological stations.

The models also include a smoothed orography that may affect comparison with data and with other models. Alterations in climate predicted within a GCM cell would likely apply fairly uniformly across regions that have relatively uniform land surface characteristics. On the other hand, steep topography or lakes smaller than GCM grids can mediate climate. Therefore, even if GCM predictions were accurate at grid scale, they would not necessarily be appropriate to local conditions (Schneider *et al.*, 1989). Since Asia's relief is varied and includes the Himalayan region, predictions of climatic change are likely to be inaccurate because models average conditions ranging from deep valleys to high mountain slopes.

Another problem arises because models differ from one another. First, model results are available on regular grids but of different sizes. This precludes a direct point-by-point intercomparison of GCM results. However, various interpolation schemes are available to alleviate this problem (Grotch, 1988). Secondly, the resolution used to represent oceans and the cloud parameterization vary between models and make comparisons between models more difficult. Finally, the effect of averaging (the last ten years of 35 GFDL and GISS year runs) model results must be carefully considered in model intercomparisons (Grotch, 1988).

5.2.4   Future climate predictions

In our study, we used $2 \times CO_2$ scenarios from simulations of the same three GCMs, the GFDL, GISS, and UKMO, supplied by the Data Support Section within the Scientific Computing Division of the National Center for Atmospheric Research (NCAR). These scenarios are based on an instantaneous equivalent doubling in atmospheric CO_2 which is then allowed to come to a new equilibrium. Equivalent doubling refers to CO_2 and other greenhouse gases having a radiative effect equivalent to the doubling of the pre-industrial value of CO_2.

The coarse grid from each model was interpolated using a four-point, inverse-distance-squared algorithm to a $0.5°$ latitude $\times$ $0.5°$ longitude grid using GRASS, a raster-based Geographic Information Systems (GIS) software package (U. S. Army Corps of Engineers, 1988) on a Sun SPARC workstation 1+. Scenarios were produced by applying *ratios* of precipitation or *differences* in temperature predicted for the $2 \times CO_2$ and $1 \times CO_2$ simulations to the baseline long-term average monthly climate dataset (Leemans & Cramer, 1992). Precipitation ratios were used to avoid negative numbers, but were not allowed to exceed 5.0 to prevent unrealistic changes in areas with normally low rainfall. This normalization assumed that, while GCMs may not reproduce the observed present-day climate very closely, the change between $1 \times CO_2$ and $2 \times CO_2$ equilibrium conditions is representative of the difference between the present climate and a future climate following an equivalent doubling of CO_2 (Parry *et al.*, 1988).

Predicted changes in annual mean temperature and annual precipitation are presented in Figs 5.1 and 5.2. All the models predict higher temperatures across the region of study. The greatest increases are predicted by UKMO. For example, in northern Japan, temperatures are predicted to change from a range of 0–10°C to one of 5–15°C, which should benefit agriculturalists in these regions. On the other hand, the current temperature range of 20–30°C over India is predicted to change to 25–35°C, which may be approaching thresholds of spikelet sterility in rice. Both GFDL and UKMO GCMs predict rainfall increases over northern India, Bangladesh, and Burma. The GFDL predicts similar increases over southeast Asia. In the same areas, the GISS predicts decreases in rainfall but significant increases over Indonesia and Malaysia.

Changes in monthly mean temperature and precipitation are presented in Fig. 5.3 for May and September, and are summarized for Asia in Table 5.3. Both GFDL and GISS predict larger temperature increases in September than in May whereas UKMO predicts larger increases in May. GFDL predicts temperature changes of 2 to 6°C in May and of 2 to 8°C in September, while GISS predicts changes in temperature ranging of 2 to 4°C in May and of 2 to 6°C in September. UKMO predicts increases in temperature of 2 to 15°C for May and of 2 to 10°C for September. Both UKMO and GISS predict an increase in precipitation (up to a doubling of current levels) in May with smaller increases in September and even a decrease in precipitation (up to 20%) over 20% of the study area. GFDL predicts an increase in precipitation over 60% of the study area and decreases elsewhere.

5.2.5 GCM limitations for future climate

The GCM predictions represent average changes that might be expected in the future following climate change. Changes in the frequency and intensity of hurricanes, the frequency of floods, and the intensity of the monsoons are much more important to the rice farmer than the average increase in monthly precipitation or temperature. However, given state-of-the-art climate models, it is still uncertain how climate variability will vary as a consequence of the increase in greenhouse gases. Few studies have analyzed the importance of climate variability to agriculture. Mearns (1991) illustrated how the CERES-wheat model responds to changes in the interannual variability of temperature in Goodland and Topeka, Kansas, USA. Panturat & Eddy (1990) studied some of the impacts of precipitation variability on rice yield in Thailand.

More importantly, there are systematic errors in simulating current climate because of weaknesses of the GCMs in representing physical processes in the atmosphere relating to clouds (Cess & Potter, 1987; Mearns, 1990) and the lack of biospheric feedback to the atmosphere (Dickinson, 1989). Even if more sophisticated models have now been designed, their predictions are not readily available outside the climatological community. Another cause of inaccuracy is the fact that we used equilibrium runs assuming an instantaneous doubling of CO_2 and a subsequent equilibrium state which, in reality, might not occur in the next 50 years if atmospheric CO_2 were to double.

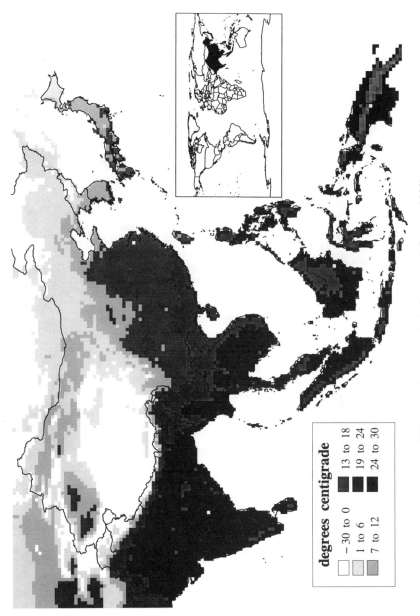

Fig. 5.1a. Current mean annual temperature (°C) across the area of study (Leemans & Cramer, 1992).

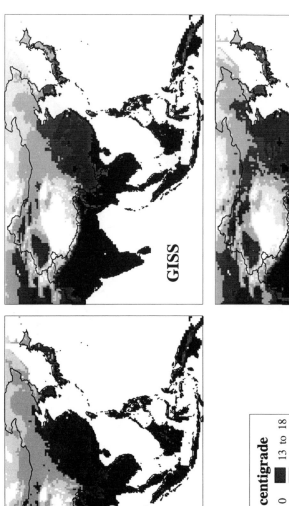

Fig. 5.1b. Mean annual temperature (°C) predicted by the three GCM 2 × CO_2 scenarios. See text for explanation.

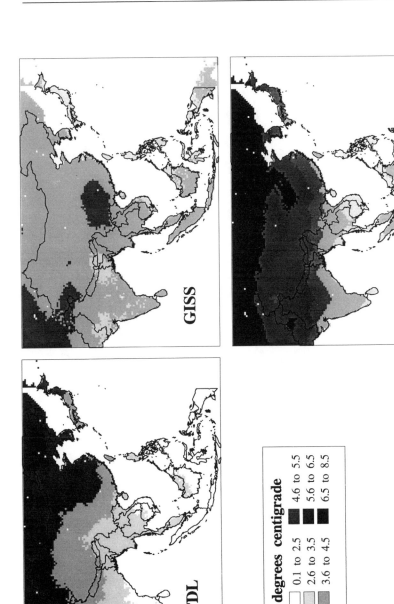

Fig. 5.1c. Mean annual temperature difference (°C) predicted by the three GCM $2 \times CO_2$ scenarios. This is the difference between Fig. 5.1a and 5.1b.

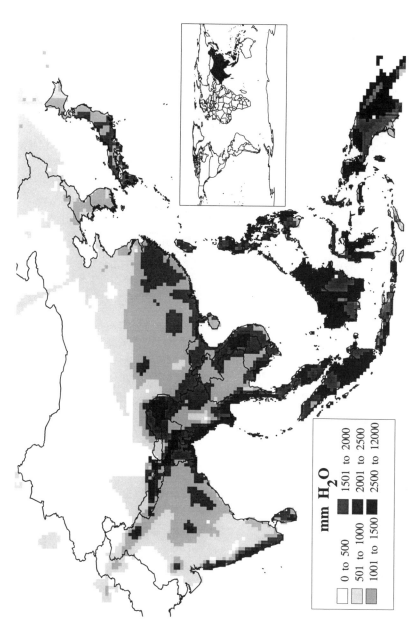

Fig. 5.2a. Current mean annual rainfall (mm) across the area of study (Leemans & Cramer, 1992).

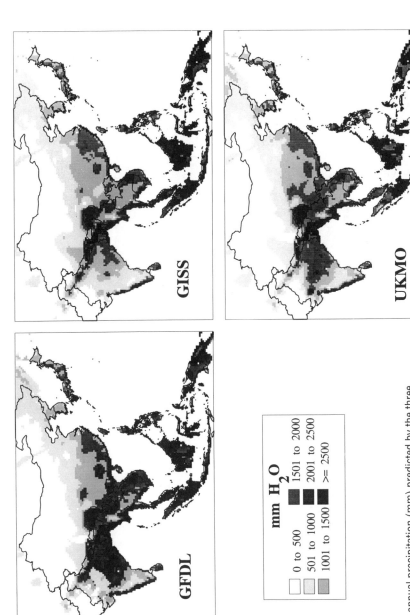

Fig. 5.2b. Mean annual precipitation (mm) predicted by the three GCM 2 × CO₂ scenarios.

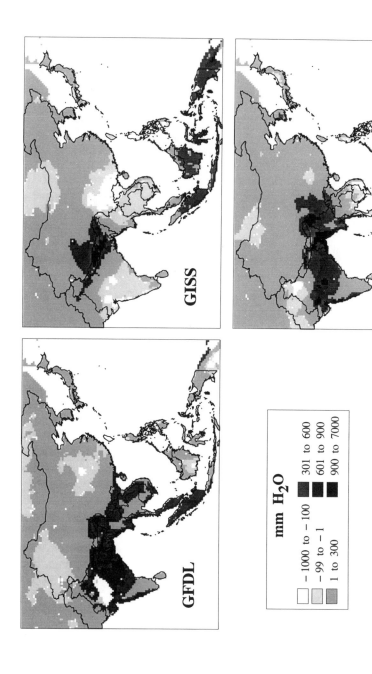

Fig. 5.2c. Mean annual rainfall difference (mm) predicted by the three GCM $2 \times CO_2$ scenarios.

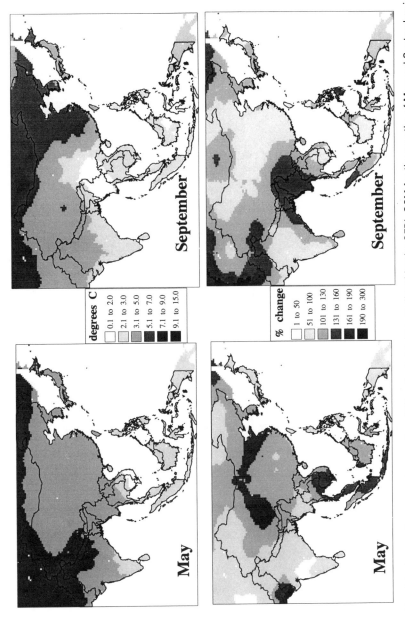

Fig. 5.3a. Monthly changes in average temperature (°C) and rainfall (%) predicted by the GFDL GCM for the months of May and September in the area of study.

D. Bachelet *et al.*

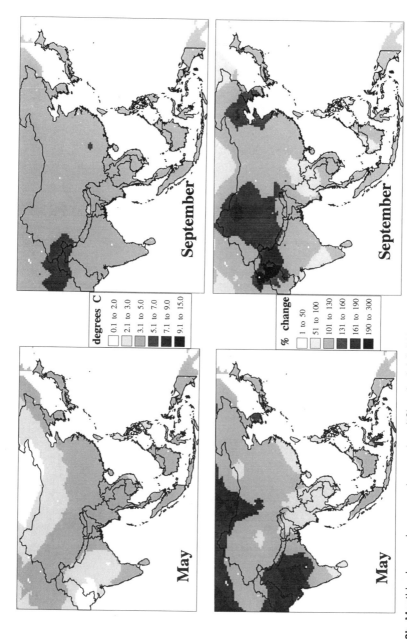

Fig. 5.3b. Monthly changes in average temperature (°C) and rainfall (%) predicted by the GISS GCM for the months of May and September in the area of study.

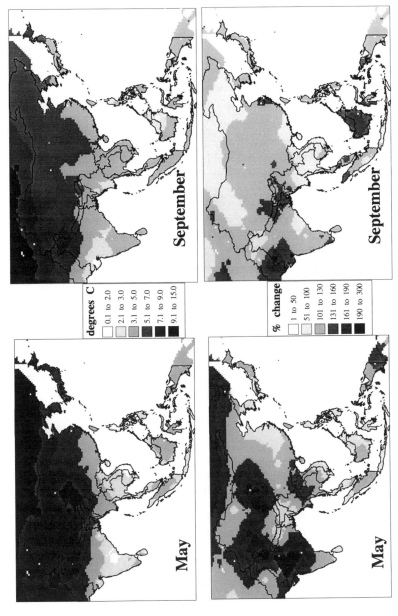

Fig. 5.3c. Monthly changes in average temperature (°C) and rainfall (%) predicted by the UKMO GCM for the months of May and September in the area of study.

Table 5.3. Fraction of the total land area subject to temperature or precipitation change according to the three GCM scenarios.

	GFDL		GISS		UKMO	
	May	Sep	May	Sep	May	Sep
Temperature (°C)						
0–2	0.04	0.02	0.19	0.00	0.01	0.00
2–4	0.33	0.27	0.73	0.47	0.17	0.19
4–6	0.48	0.39	0.08	0.53	0.13	0.27
6–8	0.15	0.28	0.00	0.00	0.28	0.35
8–10	0.00	0.05	0.00	0.00	0.22	0.17
10–15	0.00	0.00	0.00	0.00	0.19	0.01
Total	1.00	1.00	1.00	1.00	1.00	1.00
Precipitation (%)						
1–20	0.00	0.00	0.00	0.00	0.00	0.00
20–40	0.00	0.00	0.00	0.00	0.00	0.00
40–60	0.06	0.01	0.00	0.00	0.02	0.00
60–80	0.13	0.09	0.02	0.02	0.02	0.01
80–100	0.19	0.27	0.09	0.20	0.09	0.25
100–200	0.62	0.62	0.88	0.76	0.81	0.73
200–300	0.00	0.01	0.01	0.02	0.04	0.01
300–400	0.00	0.00	0.00	0.00	0.00	0.00
400–500	0.00	0.00	0.00	0.00	0.00	0.00
Total	1.00	1.00	1.00	1.00	1.00	1.00

Note: some columns do not total exactly to 1.00 due to rounding up of numbers.

5.3 Conclusions

In summary, GCMs predict a general increase of temperatures over the rice-growing regions of Asia. This can have several implications:

1. Multiple cropping would become possible where cool climate does not currently allow it. Economic incentives would greatly enhance this development.
2. Expansion of rice cultivation could take place to the north where cold temperatures currently restrict rice growth. This would only be possible if soils and management practices in these new areas were appropriate. Also population and thus urban expansion and lack of agricultural infrastructure may make it difficult.
3. A reduction of the extent of the southern rice-growing areas could be triggered if spikelet sterility would occur when high temperatures were reached or if water reserves could not meet the increased evaporation.

Of course, inherent limitations of the GCMs originally designed to study changes in the earth's temperature and not to perform regional studies should place restrictions on the extrapolations one can derive from these results. Moreover, they also cannot reliably predict changes in climatic variability such as changes in the frequency of droughts and storms which could affect productivity more than changes in average conditions. Even though its vulnerability to unfavorable climatic conditions has been demonstrated many times (e.g. Rosenberg, 1991), agriculture is expected to adjust within a few growing seasons to shifting boundaries of climatic limits to agronomic crop growth (Cramer & Solomon, 1993) by harvesting crops in newly available regions in which those crops have never been successful (Decker *et al.*, 1988; Parry & Carter, 1988), by adapting management practices (irrigation, fertilization, adapted cropping calendars), or by developing new crop varieties which can thrive under previously hostile conditions (Crosson, 1989; Rosenberg, 1991). Since world population is expected to grow extremely rapidly in the next 50 years, the quantification of those possible changes is critical, in order that planners and agricultural scientists know the type of climates to expect in the next century. Better communication between climate modelers and biologists should lead to improvements of both the General Circulation and crop simulation models, so that these predictions become increasingly reliable.

The Impact of Climatic Change on   6
Agroclimatic Zones in Asia

D. Bachelet[1] and M.J. Kropff[2]

[1]ManTech Environmental Technology, Inc., US EPA Environmental
Research Laboratory, 200 SW 35th Street, Corvallis, Oregon 97333,
USA; [2]International Rice Research Institute, P.O. Box 933, 1099 Manila,
The Philippines

6.1 Introduction

Historically, most climate classifications have originated as tools to understand
regional patterns of atmospheric circulation and corresponding spatial patterns of natural vegetation and cropping systems. Climates were often classified
on the basis of vegetative boundaries such as in the Köppen system (Strahler,
1975). However, agronomists needed analytic classifications to help them optimize the use of environmental resources to produce crops needed to feed the
rapidly growing world population. For this purpose, the Food and Agriculture
Organization of the United Nations (FAO) has published results from a crop
suitability project (Anonymous, 1978; Higgins *et al.*, 1987) using the concept
of a crop-growing period (when temperature and available water are adequate
for growth) to determine agroecological zones. Each crop variety was assigned
to one or more suitable zones characterized by specific temperature and moisture thresholds, the latter being defined by the PET formulation of Penman
(1948). However, because of its demand for extensive datasets and the lack of
adequate data for Asia, we decided not to use this approach in the present
study.

Following on from the work of Oldeman (1975, 1977), Oldeman *et al.*
(1979) and Manalo (1977a, b), Huke (1982a) published a set of agroclimatic
maps of Asia illustrating the relative lengths of the wet and dry periods. These
zones use the concept of 'dry months', when potential evapotranspiration
(PET) exceeds actual evapotranspiration (AET), and 'wet months', when rainfall (PPT) exceeds 200 mm. We tried to reproduce the location of the regional
boundaries using the same criteria, but the technique used by Huke also

included subjective information from his extensive knowledge of the region that was impossible to quantify.

Recently, Leemans & Solomon (1993) developed a similar classification based on chilling (using minimum temperatures), warmth (the sum of degree-days above 5°C), and drought requirements (using the ratio between AET and PET calculated by the Priestley–Taylor water-balance model).

In this study, we defined agroclimatic zones based on the sum of daily temperatures above 0°C (sum of degree-days above 0°C) to be compatible with the traditional definition of the major Chinese cropping systems as described by Hulme *et al.* (1992). We then combined this temperature index with a simple index of the water available to the crop, based on the difference between precipitation and potential evapotranspiration (PPT−PET), assuming that water used for irrigation was strongly correlated with rainfall, and that runoff and percolation water were either negligible or accounted for by irrigation.

6.2   Agroclimatic Zones Used in This Study

6.2.1   Region of study

The region of study extends from 60°N to 20°S latitude, and from 60°E to 150°E longitude and covers an area of 31×10^{12} m^2. It includes the following countries: China, Taiwan, North and South Korea, Japan, the Philippines, Indonesia, Vietnam, Laos, Kampuchea, Thailand, Myanmar, Bangladesh, India, Sri Lanka, and Pakistan.

6.2.2   Datasets used

The current climate baseline data set we used to determine the agroclimatic zones was the 0.5° latitude × 0.5° longitude resolution long-term average monthly climate data set from Leemans & Cramer (1992). We used the three General Circulation Model (GCM) scenarios (GFDL, GISS and UKMO), described in Chapter 5, to evaluate likely future changes in the area and distribution of these agroclimatic zones.

6.2.3   Temperature thresholds

The sum of daily temperatures above 0°C ('cumulative degree-days', or °Cd) was calculated across the region of study using the same approach as Hulme *et al.* (1992). The temperature thresholds chosen were:

1. 4000°Cd, the limit between single cropping (spring wheat, corn, sorghum, and soybean) and double cropping (wheat, corn, cotton, rapeseed, and rice) area;

2. 5800°Cd, the limit between double cropping (one crop and rice) and triple cropping (two rice crops and one other); and

3. 8000°Cd which corresponds to the limit between triple cropping area (double crop of rice plus one other crop such as wheat, rapeseed, sugar cane, fruit, or vegetable) and 'hot' triple cropping area (three crops of rice).

6.2.4 Precipitation thresholds

Water constraints on rice were defined as the difference between PPT and PET which was calculated by a turbulent transfer model (Marks, 1990). The thresholds chosen were 1000 mm difference, below which wetland rice is not commonly grown, and 2000 mm difference (10 to 12 wet months), which would allow for year-round cultivation. The reason for choosing these thresholds was that the large water losses in flooded rice fields (as a result of percolation and seepage) require from 1000 to 2000 mm of rainfall (Wopereis *et al.*, 1994).

Table 6.1. Definition of the agroclimatic zones used in this study. Each zone is defined by a temperature regime calculated as the sum of degree-days above 0°C (SUMDD) over one year, using data from Leemans & Cramer (1992). It is combined with a simple index of water availability in each region, defined as the difference between PPT, using data from Leemans & Cramer (1992), and PET, calculated using a turbulent transfer model (Marks, 1990). The table shows the percentage of the total area $(31 \times 10^6 \text{ km}^2)$ included in each zone under the current climate and future climate scenarios predicted by three GCMs for a doubled-CO_2 atmosphere.

Zone	SUMDD (°Cd)	PPT-PET (mm)	Current	GFDL	GISS	UKMO
1	> 8000	< 0	12	12	14	14
2	> 8000	0 < x < 1000	5	5	7	6
3	> 8000	1000 < x < 2000	6	8	6	9
4	> 8000	> 2000	7	9	9	9
5	5800 < x < 8000	< 0	2	5	4	6
6	5800 < x < 8000	0 < x < 1000	2	4	4	4
7	5800 < x < 8000	1000 < x < 2000	4	3	2	2
8	5800 < x < 8000	> 2000	1	1	1	1
9	4000 < x < 5800	< 0	4	11	7	14
10	4000 < x < 5800	0 < x < 1000	3	3	3	3
11	4000 < x < 5800	1000 < x < 2000	2	1	1	1
12	4000 < x < 5800	> 2000	< 1	< 1	< 1	< 1
13	0 < x < 4000	< 0	20	16	16	10
14	0 < x < 4000	0 < x < 1000	30	21	25	20
15	0 < x < 4000	1000 < x < 2000	1	1	1	1
16	0 < x < 4000	> 2000	< 1	< 1	< 1	< 1

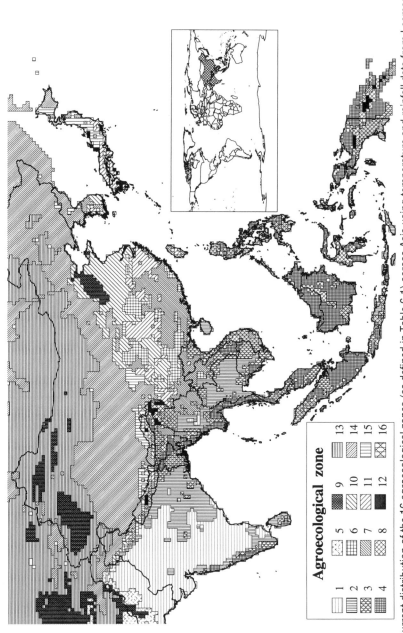

Fig. 6.1. Current distribution of the 16 agroecological zones (as defined in Table 6.1) across Asia using temperature and rainfall data from Leemans & Cramer (1992).

6.2.5 Description of the zones

Table 6.1 describes the 16 zones created using the various thresholds discussed above. Figure 6.1 shows the distribution of these zones across Asia. We then overlaid a map of the location of rice ecosystems according to Huke (1982b) to associate rice-cropping systems to climatic zones (Table 6.2). Sixty-one percent of irrigated wet rice is relatively evenly located in Zones 1, 2, 7, and 10, while 61% of irrigated dry rice is evenly located in Zones 2, 3, and 7. Irrigated hybrid rice is found in Zones 6 and 7 (55%) and 10 and 11 (42%). Rainfed (85%), deepwater (94%), and upland (76%) rice all fall mostly in Zones 2, 3, and 4, where temperature and rainfall are high.

Figure 6.2 illustrates the numbering of the agroclimatic zones and the locations of the rice ecosystems in each zone.

Table 6.2. Current distribution of rice ecosystems (as defined by Huke, 1982b) among agroclimatic zones (as defined in Table 6.1). Values represent the percentage of the total area of each ecosystem lying in each zone.

Zone	Irrigated			Rainfed			
	Wet	Dry	Hybrid	Shallow	Interm.	Deep water	Upland
1	14	6		2	8	5	10
2	15	20	1	29	42	45	35
3	11	22	1	35	31	36	24
4	4	9		17	15	13	17
5							
6	11	9	19	3	1		2
7	14	19	36	5	2		6
8		1		2	1		3
9	1		1				
10	18	5	25	2			2
11	7	8	17	3			1
12							
13	1						
14	4		1	1			1
15	1			1			
16							
Total	100	100	100	100	100	100	100

Note: some columns do not total exactly to 100 due to rounding up of numbers.

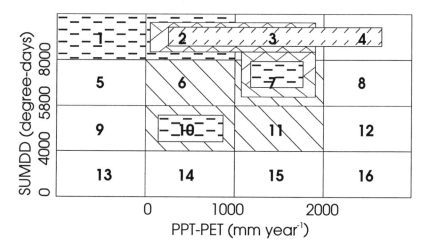

Fig. 6.2. Schematic representation of the 16 agroclimatic zones and the current distribution of rice ecosystems in these zones. Irrigated dry; irrigated wet; Upland *et al.*; hybrid.

6.3   Changes in Agroclimatic Zones

6.3.1   Climatic changes

Under all three GCM scenarios, the annual sum of degree-days was greater than that calculated for current climate conditions. The UKMO scenario was the most extreme, predicting increases greater than 1500°Cd per year over 60% of the study area. Similarly, all scenarios predicted a moderate increase in precipitation (0 to 250 mm year⁻¹) over at least 60% of the study area, with 10% of the area experiencing a more substantial increase (250 to 500 mm year⁻¹) and another 10% experiencing moderate decreases (−250 to 0 mm year⁻¹).

Under current climate conditions, annual evapotranspiration exceeds annual rainfall in 38% of the study area, but this area increased to 45%, 39%, and 44% in the GFDL, GISS, and UKMO scenarios, respectively.

6.3.2   Change in the areal extent of the zones across Asia

The most significant changes in the area of each zone were predicted to occur for Zones 5, 6, and 9, with an average increase of more than 100%, and in Zones 11 and 12, with a decrease in area of more than 60% (Table 6.3). Both Zones 6 and 11 include a significant proportion of irrigated rice ecosystems which would be affected by such areal changes.

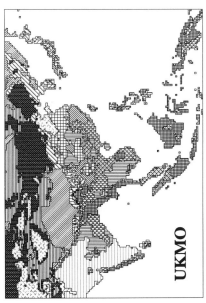
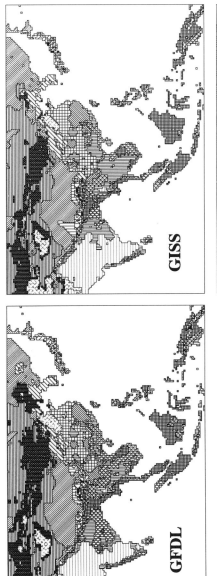

Fig. 6.3. Predicted distribution of the 16 agroecological zones across Asia under the GFDL, GISS, and UKMO 2 × CO_2 scenarios. Area extents are described in Table 6.1.

Table 6.3. Predicted change (%) in the area of the 16 agroclimatic zones (as defined in Table 6.1) under the GFDL, GISS, and UKMO scenarios.

Zone	Area ('000 km²)	GFDL	GISS	UKMO	Average change
1	3800	5	16	17	13
2	1800	-4	30	16	14
3	1800	53	11	63	42
4	2300	26	23	22	24
5	610	139	101	217	152
6	530	139	150	115	135
7	1200	-27	-40	-34	-34
8	410	-36	-52	-49	-46
9	1400	139	69	228	145
10	1100	-20	-12	0	-11
11	680	-68	-73	-73	-71
12	120	-61	-64	-78	-68
13	6400	-20	-22	-49	-30
14	9400	-28	-15	-33	-25
15	390	-42	-10	-52	-35
16	91	44	69	7	40

The relocation of the zones is illustrated in Fig. 6.3. There is a general warming of the northern regions, with many of the zones moving northwards away from the equator. Zones 1, 2, and 3 currently include a significant proportion of rice ecosystems (Table 6.2); Zone 1 was predicted to expand from the north and east of India into Pakistan and Afghanistan, Zone 2 to expand into south-eastern China, but to contract in India, while Zone 3 is expected to extend eastward in India, and northward into south-eastern China at the expense of Zones 2 and 7. Zone 7 was predicted to expand northward at the expense of Zone 11, particularly in China's Zheijiang Province, while Zone 6 was predicted to expand into eastern China at the expense of Zone 10, which in turn expands northwards at the expense of Zones 13 and 14.

6.4   Discussion

There might be significant changes in the extent and location of agroclimatic zones as we have defined them using the sum of degree-days and the difference between rainfall and PET, but they probably will not affect overall rice cultivation greatly, as already indicated by Solomon & Leemans (1990), who concluded from their studies that rice would lose 1.3% of its current cultivated area. However, since the Asian population continues to increase significantly, this reduction in cultivation area could have significant economic impacts.

The zones mostly affected by an increase in areal extent (>100%) are Zones 5, 6, and 9. Only Zone 6 (mostly Sichuan, Hunan, and Hubei Provinces of western China in Fig. 6.1) includes a significant fraction (19%) of the hybrid rice distribution and a smaller proportion of irrigated wet (11%) and dry (9%) rice ecosystems. This means that hybrid rice cultivation may be able to expand in the future if the climate evolves in the direction that the GCMs have projected.

The two zones that should see a dramatic relative decrease are Zones 11 (−67%) and 12 (−62%). Again, only Zone 11 (Zheijiang and Hunan Provinces of western China in Fig. 6.1) includes a significant fraction (17%) of the hybrid rice distribution, and a smaller proportion of irrigated wet (7%) and dry (8%) rice ecosystems.

Since the remainder of this study has focused on irrigated wet rice for which changes in potential yields were calculated, we were particularly interested in how their area of cultivation might change (Table 6.4). The most significant increase in this ecosystem was predicted to occur in Zones 2 and 3 (mostly in eastern India, Bangladesh, southern China, and the south-east Asian countries), and in Zone 6 (mostly in eastern China at the expense of

Table 6.4. Predicted distribution of the area of the wet rice ecosystem between the 16 agroclimatic zones (as defined in Table 6.1) under the three GCM climate scenarios. Values represent the percentage of the total ecosystem area contained in each agroclimatic zone.

Zone	Current	Scenarios			Average
		GFDL	GISS	UKMO	
1	14	11	16	15	14
2	15	21	22	23	22
3	11	19	14	21	18
4	4	5	5	7	6
5		2		1	1
6	11	28	24	20	24
7	14	7	11	7	8
8					
9	1	2	1	2	2
10	18	3	5	2	3
11	7		1	1	1
12					
13	1				
14	4	1	1		1
15	1				
16					

Zones 10 and 11). If, however, shifts in the distribution of this ecosystem are restricted by topography or the presence of urban or industrial centers, then temperatures should increase over most of the irrigated wet ricelands. In some areas, the effect of this in reducing crop durations may allow an extra rice crop to be grown, but in others, increased spikelet sterility may be a constraint to increased rice production, or the increased climatic variability may make such multiple cropping more precarious.

A Regional Evaluation of the Effect of Future Climate Change on Rice Production in Asia 7

R.B. Matthews[1,3], T. Horie[2], M.J. Kropff[3],
D. Bachelet[4], H.G. Centeno[3], J.C. Shin[5],
S. Mohandass[6], S. Singh[7], Zhu Defeng[8], and
Moon Hee Lee[5]

[1]Research Institute for Agrobiology and Soil Fertility, Bornsesteeg 65,
6700 AA Wageningen, The Netherlands; [2]Faculty of Agriculture, Kyoto
University, Kyoto 606, Japan; [3]International Rice Research Institute, P.O.
Box 933, 1099 Manila, The Philippines; [4]ManTech Environmental
Technology, Inc., US EPA Environmental Research Laboratory, 200 SW
35th Street, Corvallis, Oregon 97333, USA; [5]Crop Experiment Station,
Seodundong 209, Suweon, South Korea; [6]Tamil Nadu Rice Research
Institute, Adutherai, Tamil Nadu 612101, India; [7]Universiti Pertanian
Malaysia, 43400 UPM Serdang, Selangor, Malaysia; [8]China National Rice
Research Institute, 171 Tiyuchang Road, Hangzhou, 310006 Zhejiang
Province, China

7.1 Introduction

Of all the rice-producing countries in the world, rice is most closely associated
with the south, south-east, and east Asian nations extending from Pakistan to
Japan. Of 25 major rice-producing nations, 17 are located in this region, which
together produce 92%, and consume 90%, of the world's total rice production.
In 1990, the average per capita consumption of rice in Asia was 85 kg year^{-1},
providing 35% of the total caloric intake, compared with 5 kg year^{-1} (8%) in
Australia. In some Asian countries, such as Myanmar, per capita consumption
is as high as 190 kg year^{-1}, which provides some 77% of the total annual caloric
intake (IRRI, 1993). Most rice produced in Asia is consumed domestically, with
only 3–4% entering international trade markets. Major exporters are Thailand
(36%), Vietnam (10%), and Pakistan (7%), while China and India, the two
largest rice producers, export less than one million tonnes annually combined.
Because prices fluctuate more for rice than any other grain, it is regarded as a
political commodity in many Asian countries. Nevertheless, despite rapidly
expanding populations, a close balance has been maintained so far between
rice production and food needs.

Area ('000000 ha)   Average yield
or production ('000000 t)   (t ha⁻¹)

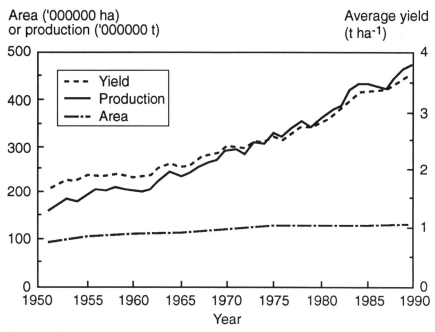

Fig. 7.1. Historical rice production statistics of Asia, showing the area sown, average national yields and total annual production. (Source: IRRI, 1991.)

Average regional trends in rice production are shown in Fig. 7.1. During the 20-year period from 1965–1967 through to 1989–1991, improvements in productivity introduced by the Green Revolution spread rapidly, so that rice production almost doubled. Most of this increase came from increased yields and cropping intensities, but some also resulted from new land brought under cultivation or shifted into rice from other crops. Much of the increase in yields was due to the introduction of semi-dwarf, short-maturing varieties, which enabled increased use of fertilizer, irrigation water, and other inputs. Further yield increases since that period have been constrained by diminishing returns and have been increasingly difficult to achieve.

However, agricultural population densities on Asia's rice-producing lands are amongst the highest in the world, and continue to increase at a remarkable rate. This rapid population growth puts increasing pressure on the already strained food-producing resources. For the leading rice-growing countries of Asia, rice requirements are predicted to increase at a compound rate of 2.1% per year (IRRI, 1993). With accelerating loss of productive ricelands to rising sea levels, salinization, erosion, and human settlements, the problem becomes one of increasing yields under even more trying circumstances. If predictions concerning population growth, climate change, and accelerating erosion of the agricultural base hold true, the turn of the century will be characterized by decreasing areas available for rice farming. Under such conditions, average yield increases of 3% per year will be needed on the remaining arable land. Such a rate of increase has never been maintained beyond a year or two for any major food crop.

7.2   The Climate of Asia

The climate of the Indian subcontinent and of east and south-east Asia is dominated by the latitudinal migration of the major windbelts following the annual course of the sun. Throughout the region, even as far north as China and Japan, precipitation patterns are strongly influenced by the southwest monsoon, which originates in the western part of the Indian Ocean at the time of high sun (Fig. 7.2b). This potentially unstable southwesterly current of maritime air crosses India, and continues eastward over the continental parts of south-east Asia and then north-eastward over much of eastern China and Japan, where it converges with the somewhat more stable, deflected North Pacific trade winds. Its passage is generally associated with the period of highest rainfall for much of south and south-east Asia, and is also a principal humidity source for the summer rainfall of eastern Asia. During the winter months, the flow reverses, and cooler air, originating in the Baikal area in Russia, flows southward, but by the time it reaches Japan and the Indo-China peninsula, it is deflected westwards towards India by the equatorial easterly trade winds (Fig. 7.2a). The convergence zone of these two flows is one of the world's most active regions of cyclogenesis, although to a much lesser extent in its westernmost parts.

The south-east Asian region (including Indonesia, Malaysia, and the Philippines) has a more complex climate due to its position on the boundaries of these major air flows, and because of the fragmented and insular nature of the land areas with the presence of numerous highlands. The area is dominated by three main air streams. To the north-east are the North Pacific trades, supplemented by the northeast monsoon in the cooler months, and to the south-east are the South Pacific trades. In the eastern part of the region, these two trades meet along the Inter-Tropical Convergence Zone, but further west, over Malaysia and Vietnam, they are separated by a wedge of air from the southwesterly monsoon described above, resulting in two more zones of convergence. Much of the weather of the region is generated along these three convergence zones, which shift northwards and southwards during the year with the sun. There are, however, numerous local perturbations caused by islands, peninsulas, and terrain barriers, which may disrupt a main air flow, resulting in localized climatic peculiarities. Because the south-west monsoon is highly humidified, and the easterlies much less so, a conspicuous feature in most of the area is that the western sides of elevated islands and peninsulas are wetter than the eastern sides, although in the case of the Malay Peninsula, where the western side lies in the lee of Sumatra, the opposite is true. A similar situation exists

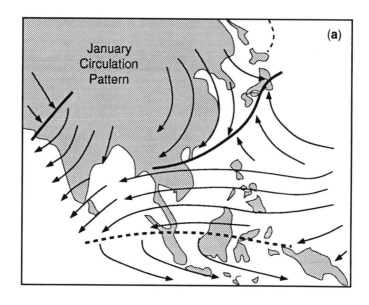

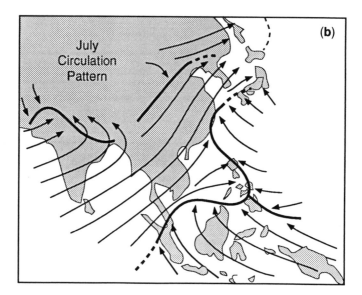

Fig. 7.2. Major air flows (arrows) dominating climate in Asia during (a) the low-sun period (January), and (b) the high-sun period (July). Solid lines represent convergence zones of major weather systems.

in the Philippines, probably because they are too far east to feel the effects of the westerly air stream for any lengthy period, instead being dominated by easterly airstreams from the Pacific Ocean. Another peculiarity worth mentioning is the small amount of annual rainfall (1000–1500 mm) and the pronounced long dry season on the southern-most islands of Indonesia, including eastern Java, southernmost New Guinea and the islands lying between them. This is mainly due to the prevalence of a relatively stable dry easterly air current originating east of Australia separately from the South Pacific trades, which prevails in the Java region for seven months of the year before giving way to the more unstable and wetter south-westerlies.

In eastern Asia, including Japan, Korea, and China, climate is characterized by seasonality, interannual variability, and sharp regionality. In the winter months, the weather is dominated by cold air flows originating from an anticyclone over eastern Siberia, and flowing south and eastwards. These flows are complex, however. Air reaching northern China has come over land, and hence is dry and cold, while that moving eastward off the continent gains moisture from the Sea of Japan, particularly in its lower layers. In southern China, air from both of these routes converge, resulting in gloomy weather with light rain and occasional heavy showers. In the summer months, three major air streams dominate; the mid-latitude zonal westerlies, the equatorial south-westerlies, and the tropical easterlies. The south-westerlies are the extension of the same south-westerly monsoon that dominates tropical south Asia in summer, which also provides the majority of moisture for the summer rainfall in east Asia. The other main source of moisture in this region is from the tropical easterlies, essentially northern tradewinds deflected by the North Pacific anticyclone, which gain moisture from the Pacific Ocean and converge with the south-west monsoon in south-east China and southern Japan. This convergence results in a region with an extremely high frequency of tropical storms and typhoons of varying intensity which can cause considerable damage. Any significant changes in the frequency, intensity, or preferred courses of tropical cyclones in response to global warming may lead to some of the most important regional impacts of climate change.

Throughout the region, much of the present interannual variability in precipitation is linked to the El Niño-Southern Oscillation (ENSO) phenomenon, which is associated with anomalies in atmospheric and oceanic circulation across the tropical Pacific and Indian Oceans (Nicholls, 1990). During the El Niño, there is a band of warm water across the equatorial central and eastern Pacific, and unusually high pressure in the Australian/Indonesian region, and with weak trade winds in the equatorial Pacific. Widespread droughts in many parts of the south and south-east Asian region are generally associated with El Niño events. Reverse anomalies occur in 'La Niña' years, with wet conditions in the south and south-east Asian regions. Some local reversals of the rainfall anomalies occur in parts of southern India, Sri Lanka, and the Indo-China region. As a consequence, any major changes in ENSO behavior under enhanced greenhouse conditions would have considerable impacts.

7.3   Rice Ecosystems in Asia

7.3.1   The irrigated rice ecosystem

Irrigated rice is grown in bunded, puddled fields with assured irrigation for one or more crops a year. This ecosystem, which is concentrated mostly in the humid and subhumid subtropics, accounts for about 80 million ha (55%) of the harvested rice area and contributes 76% of global rice production, with average yields varying from 3 to 9 t ha^{-1} (IRRI, 1993). Improved rice cultivars that have been developed for irrigated ricelands are of short duration and are responsive to nitrogen fertilizer, and incorporate resistance to several biological stresses and some tolerance to adverse soils. The ecosystem is gradually losing land to urbanization and industrialization, and farm yields are approaching the ceiling of average yields obtained in experimental stations. It is characterized by high cropping intensity and intensive use of agrochemicals.

7.3.2   The rainfed lowland rice ecosystem

Rainfed lowland rice grows in bunded fields that are flooded for at least part of the cropping season to water depths that do not exceed 50 cm for more than ten consecutive days. Rainfed lowlands are characterized by lack of water control, with floods and drought being potential problems. About 40 million ha (25%) of the total global rice area is rainfed, contributing 18% of the total production (IRRI, 1993). A proportion of the rainfed areas are favorable to rice production, and can be viewed as similar to the irrigated rice ecosystem in terms of applicable technologies and average yields. In the more unfavorable rainfed areas, yields are constrained by droughts, floods, pests, weeds, and soil constraints. Rainfall is often erratic, so that conditions are diverse and unpredictable. Typically, traditional photoperiod-sensitive cultivars which do not respond to fertilizer, are grown.

7.3.3   The upland rice ecosystem

Of the 148 million ha total rice area (1991) about 12% was planted to upland rice, with about 61% of this figure in Asia (IRRI, 1993). However, the area of the upland rice ecosystem is much larger than the area under rice, because rice is grown in rotation with other crops. Yields are typically 1 t ha^{-1}. In Asia, most upland rice is grown on rolling and mountainous land with slopes varying from 0% to more than 30%. Dry soil preparation and direct seeding in fields that are generally unbunded are characteristic of the upland rice ecosystem. Surface water does not accumulate for any significant time, although in some countries, such as in eastern India and Bangladesh, upland rice fields are frequently bunded to save scarce water. Depending on the size of their farms and their resources, upland rice farmers use cropping systems ranging from shifting to permanent cultivation. Shifting cultivation is common in Indonesia, Laos, the Philippines, northern Thailand, and Vietnam, where farmers use an area for one to three years until the fertility declines, then abandon the land, and either return to previously abandoned sites or begin cultivating new areas. Permanent cultivation, practiced in many countries, is characterized by more orderly intercropping, relay cropping, or sequential cropping with a number of crops. Inputs are often used, but because of the cost, are often below the recommended rates.

7.3.4   The flood-prone rice ecosystem

Around 10 million ha of ricelands in Asia are subject to uncontrolled flooding, mostly in the backswamps of the floodplains and deltas and on the slopes of natural levees (IRRI, 1993). The depth of water can vary from 50 cm to more than 8 m where floating rice is grown. The rice may be submerged for up to ten days, or may be subjected to long periods (1–5 months) of standing, stagnant water, or to daily tidal fluctuations causing complete submergence. Yields are low, ranging from 0.5 to 3.5 t ha^{-1}, and are extremely variable because of problem soils and unpredictable combinations of drought and flood. Floating rice yields range from 0.5 to about 2.5 t ha^{-1}, while deepwater rices may produce up to 4 t ha^{-1}. Despite these low yields, these flood-prone areas support over 100 million people.

7.4   Methods

7.4.1   Crop growth models

Two rice potential production models, ORYZA1 (v1.22) and SIMRIW, described in Chapters 3 and 4, were used to evaluate the effect of changes in temperature, solar radiation, and ambient CO_2 level on the potential production of rice in the south-east Asian region (Matthews *et al.*, 1995a). Both models, therefore, are capable of simulating production in optimally managed irrigated rice paddies, and not rainfed or upland rice fields, as water balance routines are not included in either model. The irrigated rice ecosystem represents about 75% of total world rice production (IRRI, 1993).

The main differences between the two models are described as follows:

1. *Dry matter production.* The ORYZA1 model calculates the instantaneous rate of CO_2 assimilation on an individual leaf basis from incoming solar radiation and temperature and leaf nitrogen content. It is assumed that solar radiation varies sinusoidally over the day, and that light is distributed exponentially in the canopy profile. Integration of these values both over the whole canopy and over the whole day give the gross CO_2 assimilation rate. Respiration requirements are then subtracted to obtain the daily net growth rate. In SIMRIW, on the other hand, it is assumed that daily dry matter production is proportional to the daily amount of photosynthetically active radiation (PAR) absorbed by the canopy. The proportionality constant, or radiation use efficiency (RUE), was found to be constant until the middle of the ripening stage, decreasing curvilinearly thereafter (Horie & Sakuratani, 1985). Respiration is not explicitly taken into account; instead, it is assumed that RUE integrates both CO_2 assimilation and respiration. It is further assumed that RUE is not affected by climatic conditions across a wide range of environments. In ORYZA, however, both temperature and the level of solar radiation influence the relationship between absorbed PAR and dry matter production, and respiration, which is strongly influenced by temperature, is modeled explicitly. Thus, at high and low temperatures, and low light levels, it might be expected that the two models deviate from each other to some extent.

2. *Effect of increased CO_2 level.* Figure 7.3 shows the effect of CO_2 level on crop growth rate (CGR) as simulated by the models, assuming all other factors are kept constant. The relationship used in ORYZA1 is more curvilinear than that

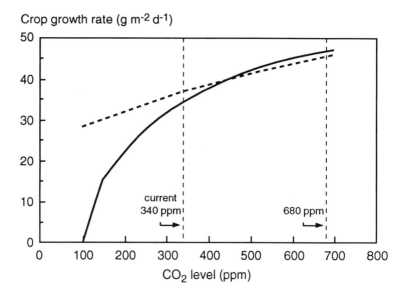

Fig. 7.3. The relation between atmospheric CO_2 concentration (ppm) and crop growth rate (g m^{-2} d^{-1}) for the two crop growth models, ORYZA1 (solid line) and SIMRIW (dashed line).

used in SIMRIW, so that ORYZA1 predicts a 7% lower CGR at the current level of CO_2 than SIMRIW, but a 3% higher value at 680 ppm. While these differences are small in themselves, the relative change in predicted yields for a doubling of CO_2 level from 340 to 680 ppm is 34% for ORYZA1 and 21% for SIMRIW. This difference between the two models will have implications for the changes in yield predicted by each in the various scenarios. Kimball (1983), summarizing data from several studies, found a 30% increase in growth rate with a doubling of CO_2 level.

3. *Leaf area development.* In ORYZA1, leaf area development is divided into three main stages: the first is during the exponential expansion stage when temperature is the main driving variable, assumed to occur when the leaf area index (LAI) is less than 1.0; the second is the linear phase of expansion, when self-shading by successive layers of leaves restricts the amount of carbohydrate available for leaf growth; and the third is the senescence period after flowering. During the second stage, leaf area growth is calculated from leaf weight growth using the specific leaf weight (SLA). Because of a 'feed-forward' loop, where small errors in leaf weight calculations are carried forward into the leaf area calculation and then back again into the leaf weight routine, this approach is very sensitive to errors in SLA. Fortunately, however, the rate of dry matter production is not very sensitive to errors in LAI calculation in this stage, due to the canopy already absorbing near the maximum fraction of incoming solar radiation. SIMRIW avoids this 'feed-forward' loop by decoupling leaf area and leaf weight calculations, and, indeed, simulates only leaf area and not leaf weight, using a series of equations specific to various growth stages similar to those used in ORYZA1. In SIMRIW, however, the only driving variable for leaf area development throughout the season is temperature, and effects of assimilate supply are not considered. Increased CO_2 levels, therefore, have no direct effect on leaf area growth. Thus, both approaches have limitations, which should be borne in mind in the following comparisons.

4. *Phenology.* In both models, phenological development rate is tracked as a function of ambient daily mean temperature and photoperiod. In ORYZA1, however, it is assumed that at temperatures above an optimum, the rate of development decreases. This assumption is based on limited data for rice, but is found to occur in a number of other crops (e.g. cassava, Matthews & Hunt, 1994; groundnut, millet, Mohamed *et al.*, 1988; sorghum, Hammer *et al.*, 1989; also Hodges, 1991). Phytotron data from IRRI indicate that a similar trend is found for rice. In SIMRIW, on the other hand, it is assumed that there is no decrease in developmental rate once the optimum temperature is reached, rather that there is a leveling off. Differences between the two models, therefore, are expected in temperature regimes above 30°C, the optimum temperature in both models. Clearly, further experimental work is necessary to determine which of the two approaches simulates reality more accurately.

5. *Dry matter partitioning.* ORYZA1 uses a table of partitioning fractions expressed as a function of the developmental stage to determine the proportions of newly produced assimilate that are allocated to the leaves, stems, roots, and panicles. The panicle dry weight may be limited by the number of spikelets formed during the period from panicle initiation to flowering, and by the fraction of these that are fertile and form grains. SIMRIW, on the other hand, only calculates the total dry weight of the crop and not the weights of individual organs during the season, and multiplies the end of season dry weight by the harvest index calculated as a function of temperature. The functions that calculate spikelet fertility and harvest index as a function of temperature are identical in both models.

ORYZA1 begins each simulation at sowing, and therefore simulates growth in the seedbed prior to transplanting, whereas SIMRIW begins simulations at transplanting. Both models were run using the GUMAYA v3.1 crop model driver (Matthews and Hunt, 1993), which can interface to a number of crop models of different types, allows multiple simulations, provides automatic determination of years available for a given weather station, error trapping of aborted runs, circular use of the same weather file if required, and can be operated in either interactive pull-down menu mode or batch mode for automated running. In addition, weather data can be either in DSSAT v3.0 (Hunt *et al.*, 1994) or in FSE (van Kraalingen *et al.*, 1991) formats; determination of the format is automatic. A termination code was included in the record of each simulation run; only runs that terminated normally or by crop death as a result of low temperatures (zero yield) were included in the analysis. Simulations terminated by insufficient weather data or other causes were not included. Weather data reading routines feature a ten-day buffer; interpolation occurred if there were missing values within this buffer only if it contained more than two real values; otherwise, the current simulation was halted and the next one in sequence was started. Input files were in the standard DSSAT v3.0 format (Hunt *et al.*, 1994), and were identical for both models, eliminating any possible source of variation due to different input data.

7.4.2 Model parameterization

Although ORYZA1 was developed using data from *indica* varieties grown at IRRI, and SIMRIW using data for *japonica* varieties growing in Japan, both models can be calibrated for different varieties by altering the appropriate input parameters, or 'genotype coefficients'. *Indica* varieties are usually grown in the tropical regions and *japonica* varieties in Japan, South Korea, and northern China; to take this into account, therefore, both models were parameterized for the *indica* variety IR64 and the *japonica* variety Ishikawi. Considerable numbers of experiments at IRRI have used IR64, providing some confidence in the parameter estimates. IR64 is early-maturing (crop duration: 83 Dd[1]), and weakly photoperiod-sensitive, similar in many respects to IR72, the most widely grown modern variety, used in testing the model in Section 3.1. Ishikawi is a very early-maturing, photoperiod-insensitive variety, developed from selections that have been made over 100 years of rice breeding in the Hokkaido region (Horie, 1988). Along with a shorter total crop duration (77 Dd) than IR64, the variety is characterized by a shorter period to flowering and a longer grain-filling duration, a pattern of development which enables the flowering period to avoid possible spikelet infertility due to the cooler temperatures later in the season to which IR64 was found to be susceptible. Values of the genotype coefficients used in ORYZA1 and SIMRIW for each of the two varieties IR64 and Ishikari are shown in Tables 7.1 and 7.2.

Table 7.1.   Genotype coefficients and function tables (expressed as a function of developmental stage, DVS) for the two varieties, IR64 and Ishikari, used in the ORYZA1 simulations. FSHTB, FLVTB, and FSTTB are the fractions of new assimilate partitioned to the shoot, leaves, and stem respectively; SLATB is the specific leaf area (cm g^{-1}), DRLVT is the relative death rate of the leaves (g g^{-1}), and NFLVTB is the leaf nitrogen concentration (g N m^{-2} leaf).

Genotype coefficient	Variable name	Units	Genotype	
			IR64	Ishikari
Initial relative leaf area growth rate	RGRL	(°Cd)$^{-1}$	0.00901	0.00901
Stem reserves fraction	FSTR	–	0.4	0.4
Basic vegetative period duration	JUDD	Dd	19.5	9.0
Photoperiod sensitive phase duration	PIDD	Dd	15.0	15.0
Minimum optimum photoperiod	MOPP	h	11.7	–
Photoperiod sensitivity	PPSE	Dd h^{-1}	0.17	0.00
Duration of panicle formation phase	REDD	Dd	25.0	20.0
Duration of grain-filling phase	GFDD	Dd	23.0	33.9
Maximum grain weight	WGRMX	mg grain^{-1}	22.8	22.8
Spikelet growth factor	SPGF	sp g^{-1}	49.2	49.2

FSHTB = 0.0, 0.5, 0.43, 0.75, 1.0, 1.0, 2.1, 1.0
FLVTB = 0.0, 0.56, 0.2, 0.56, 0.7, 0.5, 0.9, 0.3, 1.0, 0, 2, 0
FSTTB = 0.0, 0.44, 0.2, 0.44, 0.7, 0.5, 0.9, 0.5, 1.2, 0, 2, 0
SLATB = 0.000, 470, 0.152, 470, 0.336, 330, 0.653, 280, 0.787, 210, 1.011, 190, 1.431,
          170, 2.011, 170
DRLVT = 0, 0, 0.6, 0, 1, 0.015, 1.6, 0.025, 2.1, 0.05
NFLVTB = 0.000, 0.200, 0.157, 0.542, 0.333, 1.530, 0.650, 1.221, 0.787, 1.556, 1.000,
          1.288, 1.458, 1.373, 2.000, 0.836, 2.100, 0.737

---

[1] Dd represents a 'development day', which is numerically equivalent to the number of chronological days it takes to reach a given phenological event when temperature, photoperiod, and other factors are at their optimum levels. It is assumed that the number of development days required to reach each phenological stage is constant for a particular variety, although chronological days may vary depending on the ambient temperature and photoperiod (for details see Matthews & Hunt, 1994).

Table 7.2. Genotype coefficients for the two varieties, IR64 and Ishikari, used in the SIMRIW simulations.

Genotype coefficient	Variable name	Units	Genotype IR64	Genotype Ishikari
Phenology parameters: vegetative phase	GV	d	51.03	68.50
	ALF	–	0.29	0.42
	TH	°C	17.6	17.3
	BDL	–	0.66	0.47
	LC	h	23.55	16.80
	DVSAS	–	1.00	0.04
Phenology parameters: reproductive phase	TCR	°C	6.4	12.0
	GR	d	30.7	25.3
	KCR	–	0.071	0.09
Light extinction coefficient	EXTC	–	0.60	0.57
Radiation use efficiency (RUE)	COVES	g MJ^{-1}	1.95	2.00
Maximum harvest index	HIMX	–	0.42	0.37
Critical crop death temperature	CTR	°C	10.0	13.0
Leaf area production parameters:	A	d^{-1}	0.255	0.240
	KF	–	0.07	0.07
	ETA	–	0.723	0.720
	FAS	–	5.5	5.5
	TCF	°C	10.2	11.5
	BETA	–	0.5	0.5
Parameters describing cool temperature effect on spikelet fertility	THOT	°C	22.0	22.0
	STO	%	4.6	4.6
	BST	–	0.054	0.050
	PST	–	1.56	1.56

It was found that even Ishikawi could not be grown by the models for several of the years for which weather data was available at the four sites Akita, Sapporo, Sendai, and Toyama in northern Japan and Shenyang in northern China. In these areas, farmers commonly raise their rice seedlings under transparent plastic covers or in greenhouses (Horie, pers. comm.; Yoshida, 1981) where the mean temperature is typically between 21–28°C while outside ambient temperatures are still as low as 14–16°C. Transplanting then takes place when the seedlings are around 30–35 days old in May–June as temperatures begin to rise. In the models, therefore, it was assumed for these sites that the

Table 7.3. Details of weather stations used in the study, including their AEZs, and the sowing dates (SOW, day of the year) and transplanting age (TPLT, days after sowing) used in the simulations.

No.	Country	Station	Years	Longitude (°E)	Latitude (°N)	Elevation (m)	FAO-AEZ	Main season SOW	Main season TPLT	Second season SOW	Second season TPLT	Third season SOW	Third season TPLT
1	Bangladesh	Khulna	80–90	89.57	22.85	5	3	105	25	160	30	350	25
2	Bangladesh	Jessore	80–90	89.22	23.22	7	3	105	25	160	30	350	25
3	Bangladesh	Rangpur	80–90	89.25	25.75	32	3	105	25	160	30	350	25
4	Bangladesh	Joydebpur	83–92	90.43	23.90	8	3	105	25	160	30	350	25
5	China	Beijing	80–88	116.47	39.93	55	5	121	35				
6	China	Chansha	80–88	113.07	28.20	45	7	90	30	196	25		
7	China	Chendu	80–86, 88	104.02	30.67	506	6	91	40				
8	China	Fuzhou	80–88	119.28	26.08	85	7	86	30	176	25		
9	China	Guangzhou	80–88	113.32	23.13	18	7	79	30	196	25		
10	China	Guiyang	80–88	106.72	26.58	1074	6	107	40				
11	China	Hangzhou	80–88	120.20	30.23	45	7	90	30	175	25		
12	China	Nanjing	80–88	118.78	32.05	9	7	126	41				
13	China	Wuhan	80–86, 88	114.07	30.97	23	7	85	30	175	25		
14	China	Shenyang	80–88	123.43	41.77	42	5	121	35				
15	India	Aduthurai	60–65, 68–92	79.50	11.00	19	1	161	25	350	25		
16	India	Coimbatore	83–85, 88–90	77.00	11.03	431	1	161	25	350	25		
17	India	Cuttack	83–87, 90	86.00	20.50	23	2	213	25	15	25		
18	India	Hyderabad	83–84	78.43	17.42	545	1	160	25	350	25		
19	India	Kapurthala	83–85	75.87	30.93	247	6	191	25				
20	India	Patancheru	75–84	78.35	17.50	25	1	160	25	350	25		
21	India	Bijapur	72–80	75.80	16.78	594	1	160	25	350	25		
22	India	Madurai	86–91	78.02	8.50	147	1	161	25	350	25		

Continued

Table 7.3. *Continued.*

No.	Country	Station	Years	Longitude (°E)	Latitude (°N)	Elevation (m)	FAO–AEZ	Main season SOW	Main season TPLT	Second season SOW	Second season TPLT	Third season SOW	Third season TPLT
23	India	Pattambi	83–85	76.20	10.80	25	2	120	25	300	25		
24	Indonesia	Bandung	80–86, 88	107.52	–6.92	791	3	270	20	80			
25	Indonesia	Ciledug	81–88	106.75	–6.25	26	3	270	20	80			
26	Indonesia	Cimanggu	89–92	106.73	–6.62	240	3	270	20	80			
27	Indonesia	Cipanas	89–91	107.02	–6.75	1100	3	270	20	80			
28	Indonesia	Maros	75–89	119.50	–5.00	5	3	270	20	140			
29	Indonesia	Muara	81–85, 90–92	106.67	–6.75	260	3	270	20	80			
30	Indonesia	Pacet	89–92	107.03	–6.75	1138	3	270	20	80			
31	Japan	Akita	79–90	140.10	39.72	9	8	107	31				
32	Japan	Hiroshima	79–90	132.43	34.37	29	8	113	24				
33	Japan	Kochi	79–90	133.53	33.55	2	8	134	26				
34	Japan	Miyazaki	79–90	131.42	31.92	7	8	134	26				
35	Japan	Maebashi	79–90	139.07	36.40	112	8	139	30				
36	Japan	Nagoya	79–90	136.97	35.17	51	8	121	25				
37	Japan	Sapporo	79–90	141.33	43.05	17	8	108	36				
38	Japan	Sendai	79–90	140.90	38.27	39	8	96	31				
39	Japan	Toyama	79–90	137.20	36.70	9	8	99	25				
40	Malaysia	Kemubu	83–90	102.33	5.93	1	3	221	25	69	25		
41	Malaysia	T. Chengai	78–88	100.28	6.10	1	3	221	25	69	25		
42	Malaysia	Tanj. Karang	80–90	101.20	3.50	2	3	221	25	69	25		
43	Myanmar	Myananda	85–90	98.42	17.50	–	2	140	20	290	20		
44	Myanmar	Pyay	85–90	97.00	18.00	–	2	140	20	290	20		
45	Myanmar	Wakema	85–90	95.65	16.83	–	2	140	20	290	20		
46	Myanmar	Yezin	83–90	96.07	21.95	74	2	140	20	290	20		

#	Country	Station	Years										
47	Philippines	Albay	72–73, 75–89	123.70	13.38	—	3	135	20	15	20	258	20
48	Philippines	Butuan	81–90	125.73	11.03	60	3	135	20	15	20	258	20
49	Philippines	CLSU	74–90	120.90	15.72	76	3	135	20	15	20	258	20
50	Philippines	CSAC	75–90	123.27	13.57	36	3	135	20	15	20	258	20
51	Philippines	La Granja	75–90	122.93	10.40	84	3	135	20	15	20	258	20
52	Philippines	MMSU	76–88	124.30	8.00	854	3	135	20	15	20	258	20
53	Philippines	PNAC	77–81, 83–90	118.55	9.43	7	3	135	20	15	20	258	20
54	Philippines	UPLB	59–90	121.25	14.17	21	3	135	20	15	20	258	20
55	Philippines	IRRI Wetland	79–92	121.25	14.18	21	3	135	20	15	20	258	20
56	South Korea	Cheongju	77–90, 91–92	127.43	36.63	59	6	105	40				
57	South Korea	Chilgok	77–84, 86–89	128.57	35.95	55	6	110	40				
58	South Korea	Chinju	77–89, 91–92	128.10	35.20	22	6	110	40				
59	South Korea	Chuncheon	77–88, 91–92	127.73	37.90	74	6	105	40				
60	South Korea	Taejeon	77–88, 91–92	127.40	36.30	77	6	110	40				
61	South Korea	Kangneng	77–89, 91–92	128.90	37.75	26	6	95	40				
62	South Korea	Kwangju	77–88, 91–92	126.92	35.13	71	6	110	40				
63	South Korea	Cheonju	77–87, 91–92	127.15	35.82	51	6	110	40				
64	South Korea	Milyang	73–88, 91–92	128.73	35.48	12	6	110	40				
65	South Korea	Suweon	77–89, 91–92	126.98	37.27	37	6	105	40				
66	South Korea	Cheolweon	77–88	127.13	38.15	192	6	95	40				
67	Taiwan	Pingtung	83–90	120.50	22.67	24	7	350	20	120	20		
68	Thailand	Chiang Mai	75–88	98.98	18.78	313	2	150	28	350	20		
69	Thailand	Khon Kaen	75–88	102.83	16.43	165	2	180	28	350	20		
70	Thailand	Ubon Ratchathani	75–90	104.87	15.25	123	2	180	28	350	20		

mean temperature was 24°C and the daily solar radiation 70% of the observed value until transplanting, after which the observed values only were used.

It is recognized that both cultivars may differ from those actually used in each region, particularly in phenology, so that the potential yields simulated in this study may not be the same as those simulated using parameters for the local varieties. Nevertheless, most irrigated high-yielding varieties (HYVs) are based on IRRI genotypes similar to IR64; it is assumed, therefore, that the trends in the simulated effects of climatic change are valid for most varieties. It is the purpose of the case studies for individual countries later in this book to evaluate more precisely the effect of climate change in particular regions using the model parameterized for local cultivars.

7.4.3   Weather stations

Weather data from 68 stations from 11 countries in the south-east Asian region were used in the analysis (Table 7.3). Fuller details of the stations, including their location, summary statistics, and number of years' data for each, are found in Chapter 2. In total, there were about 780 years of weather data available.

Weather stations were classified according to the agroecological zone in which they were located. The zoning system was that developed by the Food and Agriculture Organization (FAO) based on climatic conditions and landforms that determine relatively homogeneous crop-growing environments. The system distinguishes between tropical regions, subtropical regions with winter or summer rainfall, and temperate regions. These major regions are further subdivided into rainfed moisture zones, lengths of the growing period, and thermal zones based on the temperature regime that prevails during the growing season (Table 7.4). Most of the countries in the study fell within a single agroecological zone (AEZ), although India and China spanned several zones.

Table 7.4.   Description of the FAO-AEZs used in the study (IRRI, 1993).

Zone	Description
1	Warm arid and semi-arid tropics
2	Warm subhumid tropics
3	Warm humid tropics
5	Warm arid and semi-arid subtropics with summer rainfall
6	Warm subhumid subtropics with summer rainfall
7	Warm/cool humid subtropics with summer rainfall
8	Cool subtropics with summer rainfall

To avoid confusion with the national AEZ systems used in the individual country chapters (Chapters 8–13) in this book, the AEZs defined by FAO and used in this and subsequent chapters are referred to as FAO-AEZs.

7.4.4   Determination of planting dates

Dates of sowing and transplanting were, in general, supplied by the collaborating institutions along with the weather data. Where this information was not given, transplanting dates were obtained from IRRI (1991), and date of sowing in the seedbed assumed to be 25 d prior to this. Where a range of transplanting dates was given, generally a date near the start of the range was used. In some countries, second, and even third, crops are grown in the same year; these were also simulated. Sowing dates and ages at transplanting for each season at each site are shown in Table 7.3. These agree well with the dates used by Jansen (1990) with the exception of those in Indonesia; both are within the range given in IRRI (1991), however.

It is recognized that these dates sometimes may be somewhat arbitrary and not always a reflection of actual planting dates in a given region. Published crop calendars for a number of regions are available, but there is often disagreement between these even for the same regions, thereby limiting their use. Transplanting dates depend on the decisions of individual farmers, which are influenced by actual weather conditions, economic considerations, and other factors. Often transplanting in a region can take place over extended periods, particularly in tropical regions; in the higher latitudes, generally there is a very constant planting date (Chapter 8).

It was assumed for all scenarios that planting dates were the same as current dates. It is recognized that farmers will probably adjust planting dates in response to a gradually changing climate, but it is a major modeling exercise to attempt to optimize planting dates at each site in addition to evaluating the direct effects of the climate change. In order for planting date optimization to be meaningful, consideration should also be taken of precipitation and water availability, which we have not done in the present analysis. Nevertheless, in a subsequent section, we have used an example to illustrate how planting dates may be changed to avoid damaging temperatures around the time of flowering.

7.4.5   Climate change scenarios

In total, the impact of 15 different climate change scenarios were evaluated. These included scenarios in which temperature and CO_2 were varied in fixed increments above the current temperatures for each site, both independently and in combination, as well as the scenarios predicted for a doubled CO_2 climate $(2 \times CO_2)$ by the three General Circulation Model (GCM) scenarios described in Chapter 5. These were the General Fluid Dynamics Laboratory model (GFDL), the Goddard Institute of Space Studies model (GISS), and the United Kingdom Meteorological Office model (UKMO). Each GCM has used a slightly different current CO_2 concentration (Table 5.1); the $2 \times CO_2$ scenario of each does not, therefore, represent the same CO_2 level. However, to allow meaningful comparisons, we have taken the baseline, or current level of CO_2 as 340 ppm, and have used a CO_2 level of twice this value (i.e. 680 ppm) for each of the GCM scenarios, so that only the differences in predicted changes in temperature and solar radiation were evaluated. It is recognized that, as the scenarios are predictions of climate under an equivalent doubling of CO_2, the actual CO_2 level associated with the scenario may be somewhat less than 680 ppm, due to the contribution to warming from other greenhouse gases. However, due to the uncertainty associated with which level of CO_2 should be used with an equivalent $2 \times CO_2$ scenario, we feel that the use of 680 ppm is valid.

Nevertheless, it is useful to have some idea of the sensitivity of the predicted changes to differences in the CO_2 level. For this, we selected three weather stations representing a northern latitude (Shenyang, China; 41.8°N), a southern latitude (Madurai, India; 8.5°N), and one midway between these extremes (Khulna, Bangladesh; 22.9°N). We then made simulations with the ORYZA1 crop simulation model for a single year at each station for each of the three GCM scenarios, firstly using a CO_2 level of 680 ppm, and secondly using a level of 560 ppm. From these simulations, we calculated the fractional change above the predicted current potential production for the nine site/GCM combinations at each CO_2 level. Figure 7.3 shows the predicted changes plotted against each other. Where large decreases are predicted, there is very little effect of changing the CO_2 level from 680 ppm to 560 ppm, but where large increases are predicted, the difference is more marked. Large decreases occur where temperatures, either high or low, are dominant, and CO_2 level is relatively unimportant. Large increases mainly occur where CO_2 level is more influential. The regression intercept indicates that around the zero-change level, there is about a 7% difference in predicted changes between using the 560 ppm and 680 ppm CO_2 levels. In the following analysis, therefore, it should be kept in mind that large predicted increases may overestimate slightly what would happen under a 'doubled-CO_2' climate.

A description of each of the 15 scenarios is given in Table 7.5. It was assumed that the pattern of day-to-day temperature changes in each scenario would be the same as for the present observed climate (represented by daily station data) but that each daily temperature will be adjusted by the appropriate monthly change predicted in the scenario. This may not be entirely true, particularly in the GCM scenarios, but until more accurate fine-scale predictions are available, it is not possible to know the error introduced by this assumption.

Simulations were made for each of the 68 weather stations using all the years of data available for each, and using the main season planting dates shown in Table 7.3. For a further 44 stations located in the double-cropped

Table 7.5. Description of the climatic change scenarios used in the study.

Scenario	Description
1	Current climate: The CO_2 level was assumed to be 340 ppm, and measured values of temperature and radiation from each weather station were used.
2	Current climate +1°C: Same as scenario 1 but with 1°C added to both max. and min. temperatures.
3	Current climate +2°C: Same as scenario 1 but with 2°C added to both max. and min. temperatures.
4	Current climate + 4°C: Same as scenario 1 but with 4°C added to both max. and min. temperatures.
5	Current climate but with 1.5 × CO_2 (510 ppm) ambient CO_2 concentration.
6	Current climate but with 2 × CO_2 (680 ppm) ambient CO_2 concentration.
7	Current climate but with 1.5 × CO_2 (510 ppm) ambient CO_2 concentration, and 1°C added to both max. and min. temperatures.
8	Current climate but with 1.5 × CO_2 (510 ppm) ambient CO_2 concentration, and 2°C added to both max. and min. temperatures.
9	Current climate but with 1.5 × CO_2 (510 ppm) ambient CO_2 concentration, and 4°C added to both max. and min. temperatures.
10	Current climate but with 2 × CO_2 (680 ppm) ambient CO_2 concentration, and 1°C added to both max. and min. temperatures.
11	Current climate but with 2 × CO_2 (680 ppm) ambient CO_2 concentration, and 2°C added to both max. and min. temperatures.
12	Current climate but with 2 × CO_2 (680 ppm) ambient CO_2 concentration, and 4°C added to both max. and min. temperatures.
13	Current climate but with adjustments to temperature and solar radiation predicted by the GFDL GCM model with 2 × CO_2 scenario.
14	Current climate but with adjustments to temperature and solar radiation predicted by the GISS GCM model with 2 × CO_2 scenario.
15	Current climate but with adjustments to temperature and solar radiation predicted by the UKMO GCM model with 2 × CO_2 scenario.

zone (Chapter 2), simulations were made for the second season, and for another 13 stations where triple-cropping is practiced, a third season. The 780 years of weather data combined with the two models, the 15 scenarios, and the multiple plantings, gave a total number of simulations of around 50,000. Running in batch mode on a 486DX/66MHz desk-top computer, these took about 17–18 h to complete.

7.5 Analysis

7.5.1 Comparison of the two crop growth models

Figure 7.5 shows a comparison of the predicted potential yields at each of the

Predicted change using 560 ppm

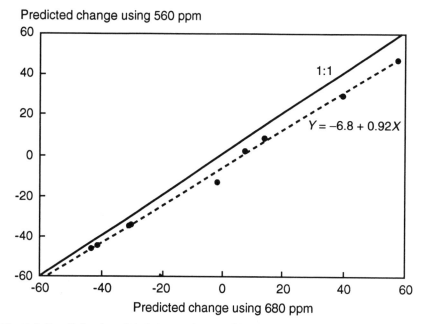

Fig. 7.4. Sensitivity of predicted changes in potential rice yields to CO_2 level. The *x*–axis represents the predicted changes in yield (%) under the three $2 \times CO_2$ GCM scenarios using a 680 ppm level of CO_2, while the *y*–axis represents the predicted changes in yield under the same $2 \times CO_2$ scenarios but using a 560 ppm level of CO_2. See text for details.

68 sites between each of the two models. Current potential yields, as well as those at the +4°C temperature increment to represent conditions under the lower end of the yield range, and at doubled-CO_2 levels to represent yields at the top of the range, are shown for the main planting season at each site. Other combinations of temperature and CO_2 levels in both the 'fixed-level' scenarios and the GCM scenarios will fall between these two extremes, and are not shown here.

In general, there was good agreement between the two models, although there were tendencies for the models to deviate at both the high and low ends of the yield range. At the low end of the range, SIMRIW predicted higher yields than ORYZA1, which is explained by the difference in the way respiration is accounted for in each model (see Section 7.4.1). SIMRIW assumes that RUE, and hence respiration, is not affected by temperature, whereas in ORYZA1, respiration is an exponential function of temperature, so that at higher temperatures, proportionately more of the assimilated CO_2 is lost from the crop by respiration. Thus, at higher temperatures, corresponding crop growth rates were lower in ORYZA1 than SIMRIW. At the high end of the yield range, there was a slight tendency for SIMRIW to predict lower potential yields than

SIMRIW yield ('000 kg ha⁻¹)

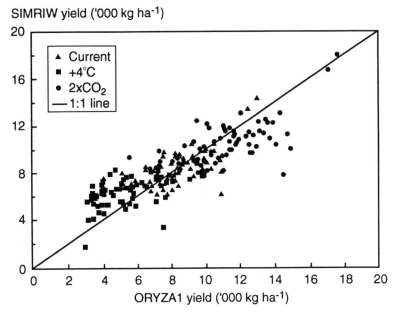

ORYZA1 yield ('000 kg ha⁻¹)

Fig. 7.5. Comparison of potential yields (kg ha⁻¹) simulated by the ORYZA1 and SIMRIW rice simulation models. Simulations were made using current weather variables, with daily mean temperature incremented by +4°C, and with doubled-CO_2. Values are the means across all years available for each weather station and for the main planting season only.

ORYZA1, which was due to the relationship between CO_2 level and RUE being lower at the 680 ppm level in SIMRIW (Fig. 7.5), and also to differences in the maximum leaf area attained.

7.5.2 Potential production under current climates

The potential yields predicted by each of the two crop simulation models are shown in Fig. 7.6. For ORYZA1, the lowest potential yield in the main planting season was 4.0 t ha⁻¹ at Nanjing in China, while the highest potential yield was 13.0 t ha⁻¹ at Kapurthala, in northern India, a site characterized by high solar radiation and low temperatures during the growing season. For SIMRIW, yields ranged from 6.3 t ha⁻¹ at Cimanggu in Indonesia to 14.5 t ha⁻¹, also for Kapurthala. Mean yields for the two models, averaged across all sites and for all available years, were 8.10 and 8.49 t ha⁻¹ for ORYZA1 and SIMRIW respectively, a difference of only 5%. These values are comparable to those obtained by Jansen (1990), although the range is somewhat larger, probably because of the larger number of sites included in our analysis compared to the seven used in that study.

Figure 7.7 shows the relationship between the latitude and main season

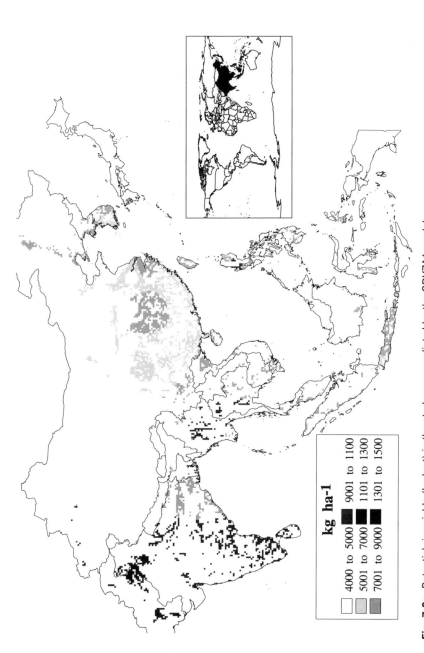

Fig. 7.6a. Potential rice yields (kg ha⁻¹) in the study area predicted by the ORYZA1 model.

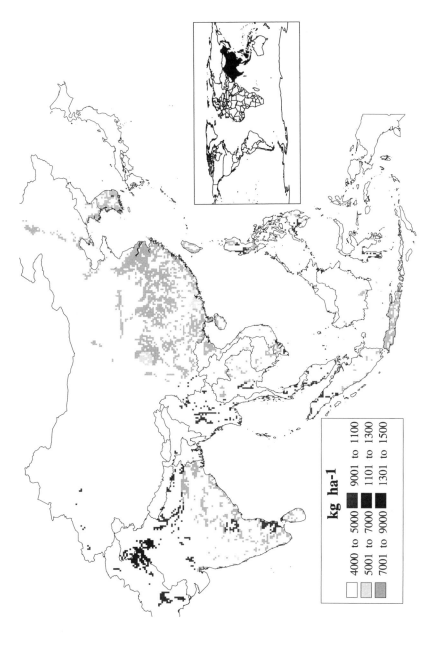

Fig. 7.6b. Potential rice yields (kg ha^{-1}) in the study area predicted by the SIMRIW model.

kg ha-1

4000 to 5000	9001 to 1100
5001 to 7000	1101 to 1300
7001 to 9000	1301 to 1500

Potential yield (kg ha⁻¹)

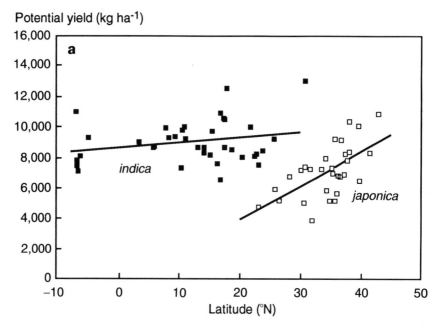

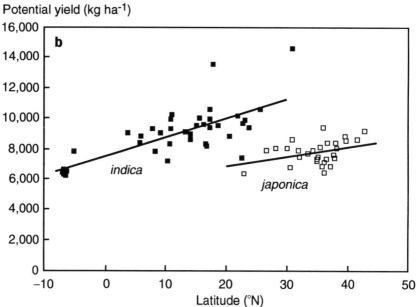

Fig. 7.7. Relationship between latitude and main season potential yield (kg ha⁻¹) for (a) the ORYZA1 model, and (b) the SIMRIW model.

potential yield predicted for each site. For both models, and within both *indica* and *japonica* ecotypes, there is a tendency for yields to be higher at higher latitudes, although it is interesting to note that the predicted maximum potential yields (10–11 t ha^{-1} for both models) of the *japonica* variety in the temperate regions (i.e. Korea, China, and Japan) are not generally higher than those of the *indica* variety in the more tropical regions, contrasting with the claims of some authors (e.g. Yoshida, 1983). Assuming the present set of weather stations adequately represents the range of rice-growing environments, the high yields that have been obtained at specific locations at the higher latitudes (e.g. Yoshida, 1981) would therefore seem to be not a general phenomenon, but rather a result of the particular combination of low temperatures and high solar radiation at those sites.

There was no significant correlation between predicted potential yields and mean temperature over the season, nor the mean temperatures over the panicle formation or grain-filling phases. However, much of the variation in yields between sites could be explained in terms of incident solar radiation (S_o, MJ m^{-2} d^{-1}); correlations between yield and S_o from panicle initiation (PI) to flowering ($r^2 = 0.307$, $n = 68$), S_o during grain-filling ($r^2 = 0.088$, $n = 68$), and S_o from PI to maturity ($r^2 = 0.204$, $n = 68$) were all highly significant ($P < 0.001$). However, an even better correlation was found between the photothermal quotient (Q, MJ m^{-2}d^{-1}°C^{-1}) and grain yield (Fig. 7.8), where Q is calculated as the ratio of S_o to the mean temperature (T_m, °C) minus the base temperature (T_b, 8°C), over the period from PI to flowering, i.e.

$$Q = S_o / (T_m - T_b) \qquad (6.1)$$

Previous work has already shown a strong relationship between Q and final grain yield in rice (e.g. Islam & Morison, 1992). Interestingly, there was a significant difference in the relationship between Q and yield in the *indica* and *japonica* varieties (Fig. 7.7), with a lower y-intercept in the latter. The variability of the regression was also greater for the *japonica* variety, indicating that other factors must be responsible for determining yield; spikelet fertility response to high and low temperatures seems a likely candidate.

It should be pointed out that the main season yields are not necessarily the highest potential yields that can be obtained at each site. In the Philippines, for example, yields obtained during the dry season are usually 30–40% higher than those obtained in the wet season, primarily due to higher solar radiation levels. However, factors other than temperature and solar radiation determine the optimum time of planting, the major one of which is rainfall through its effect on water availability.

As expected, there was a much stronger relationship between the annual potential production (the sum of all multiple-season plantings) and the latitude of each site, with the highest annual productions of nearly 30 t ha^{-1} in the 10–15° range, where three crops a year are possible (data not shown). The lower annual production at sites closer to the equator, primarily in Malaysia

Yield (kg ha^{-1})

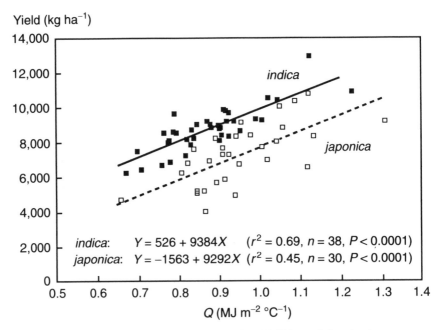

Fig. 7.8. Relationship between photothermal quotient (Q, MJ m^{-2} °C^{-1}) and main season potential yield (kg ha^{-1}) simulated by the ORYZA1 model for the *indica* variety IR64, and the *japonica* variety Ishikawi. Similar relationships were predicted by the SIMRIW model.

and Indonesia, are due to only two crops per year being grown because of the practical difficulties encountered in obtaining a third crop; heavy rains during this period make harvesting very difficult (A. Rajan, pers. comm.).

7.5.3   Effect of fixed increments of temperature and CO$_2$ on potential yields

Table 7.7 shows the overall effect predicted by both crop models of changes in temperature and CO$_2$ levels on potential yields, averaged across all sites, seasons and years. The changes predicted for each season were weighted by the factors shown in Table 7.6 for each country to account for differences in production between seasons. Both models predicted that at all CO$_2$ levels, an increase in temperature would cause a decline in yields, and at all temperature increments, an increase in CO$_2$ would increase yields. Similar patterns of interaction have also been found using other models (e.g. Bachelet *et al.*, 1993). At the current level of CO$_2$, 340 ppm, ORYZA1 predicted an average of -7.4% change in yields for every 1°C increase in temperature, while SIMRIW predicted -5.3% °C^{-1}. The higher value predicted by the SIMRIW model is due to there being no direct effect of temperature on respiration rates, and hence no

Table 7.6. Factors used to weight the predicted changes for each season to account for differences in the contribution to overall annual production.

Country	Main season	Second season	Third season
Bangladesh	0.80	0.18	0.02
China	0.66	0.34	–
India	0.80	0.20	–
Indonesia	0.80	0.20	–
Japan	1.00	–	–
Malaysia	0.50	0.50	–
Myanmar	0.80	0.20	–
Philippines	0.80	0.18	0.02
South Korea	1.00	–	–
Taiwan	0.80	0.20	–
Thailand	0.80	0.20	–

decrease in overall dry matter production at higher temperatures. The ORYZA1 value is closer to the 7–8% °C^{-1} decrease measured by Baker *et al.* (1992b) in controlled environment experiments. Further experimental work is needed to clarify the true nature of the relationship between temperature and rate of dry matter production. As might be expected from Fig. 7.3, ORYZA1 predicted larger yield increases with increases in the CO_2 level at current temperatures than did SIMRIW (Table 7.7). However, when the temperature was incremented by +4°C, these differences between the models became remarkably small.

Increases in temperature also increased the overall variability (represented by the coefficient of variation, CV%) in yields predicted by both models, more so

Table 7.7. Mean predicted changes (%) in potential yields under the 'fixed' temperature and CO_2 scenarios. Temperature increments are above the current mean temperatures at each site. Changes are averaged across all sites and all available years.

	Temperature increment			
	+0°C	+1°C	+2°C	+4°C
ORYZA1				
340 ppm	0.00	−7.25	−14.18	−31.00
1.5 × CO_2	23.31	14.29	5.60	−15.66
2 × CO_2	36.39	26.42	16.76	−6.99
SIMRIW				
340 ppm	0.00	−4.58	−9.81	−26.15
1.5 × CO_2	12.99	7.81	1.89	−16.58
2 × CO_2	23.92	18.23	11.74	−8.54

in SIMRIW than in ORYZA1 (Table 7.8). An increase in the CO_2 level had little effect on yield variability in ORYZA1, and no effect whatsoever in SIMRIW. This is a consequence of the effect of CO_2 level being accounted for in SIMRIW by a simple multiplier function which adjusts all yields from every year by a single factor, without affecting their variance. In ORYZA1, CO_2 concentration affects the two photosynthetic parameters ϵ and A_m in a non-linear way (Equation 3.8), resulting in a small effect on the variance of yields.

Table 7.8.   Mean coefficients of variation (%) for each of the 'fixed' temperature and CO_2 scenarios. Values are calculated across years for each site, then averaged across sites.

	Temperature increment			
	+0°C	+1°C	+2°C	+4°C
ORYZA1				
340 ppm	14.9	15.7	16.7	21.6
1.5 × CO_2	13.8	14.5	15.9	21.5
2 × CO_2	13.3	14.0	15.4	21.2
SIMRIW				
340 ppm	8.9	10.1	12.3	21.7
1.5 × CO_2	8.9	10.1	12.3	21.7
2 × CO_2	8.9	10.1	12.3	21.7

7.5.4   Effect of predicted GCM scenarios on potential yields

Averaged across all sites and all years and all planting seasons, the ORYZA1 model predicted overall changes of +3.6%, −3.2%, and −8.3% for the GFDL, GISS, and UKMO scenarios respectively, while the SIMRIW model predicted corresponding changes of +0.8%, −5.7%, and −10.7%. Although there were differences in the absolute values of the predicted changes of each crop model, the ranking of each of the GCM scenarios is the same; the GFDL scenario is the least, and the UKMO the most, detrimental to rice yields. This ranking corresponds to the temperature increase predicted by each GCM (Table 5.1), and is similar to that found in the study of Rosenzweig *et al.* (1993) using the CERES-Rice model (Godwin *et al.*, 1993).

However, as there are large differences in the production capacities between countries, it is more meaningful to weight the average increase by the current total rice production of each country. It is recognized that national yield figures are composed of a combination of rainfed and irrigated production, different varieties and soils, and a range of management levels, but as a first analysis, it is assumed that the proportional change in potential production is the same as that for the national production. To obtain a more realistic estimate of yield changes in India and China, these large countries were subdivided into their component FAO-AEZs. All other countries lay in single FAO-AEZs. In the case of India and China, a small amount of rice is produced in FAO-AEZ 8, but because no weather station data were available for this zone, the proportional increases for FAO-AEZ 8 of Japan were used. Similarly, no weather data were available for FAO-AEZ 5 in India, and the average of FAO-AEZ 5 in China was used. In the first case, rice production from FAO-AEZ 8 is small (0.8% of total regional production), so errors in predictions are unlikely to affect the overall prediction for the region, but in the second case, production from FAO-AEZ 6 was 6% of current total regional production, so that any errors here could influence the total slightly. Proportional changes in each country or zone were calculated on the simulated total annual production, including all plantings for the year, which were weighted according to their contribution using the values in Table 7.6. With the available data, these estimates of regional rice production are probably the most accurate it is possible to make, although it is difficult to provide estimates of the magnitudes of the errors involved.

Table 7.9. Estimated changes in total rice production predicted by the ORYZA1 model for each country and in the region under the three GCM scenarios. Current actual production ('000 t) in each AEZ on a country basis are adjusted by the simulated changes in total annual production. See text and Matthews *et al.* (1995b) for explanation.

Country	AEZ	Current[1] '000 t	GFDL % change	GFDL '000 t	GISS % change	GISS '000 t	UKMO % change	UKMO '000 t
Bangladesh	3	27,691	14.2	31,621	−5.0	26,298	−2.8	26,919
China	5	8,854	−7.4	8,201	0.3	8,881	−25.2	6,619
	6	79,872	0.8	80,484	−21.7	62,514	−19.5	64,334
	7	91,828	5.8	97,196	5.8	97,135	3.1	94,695
	8	2,361	−6.4	2,209	−14.2	2,026	−27.6	1,710
India	1	32,807	4.6	34,305	−10.8	29,272	−5.5	31,017
	2	49,949	1.8	50,849	−2.9	48,493	−7.9	46,002
	5	227	−7.4	210	0.3	228	−25.2	170
	6	26,628	5.4	28,069	3.2	27,480	−1.3	26,287
	8	1,011	−6.4	946	−14.2	867	−27.6	732
Indonesia	3	44,726	23.3	55,155	9.0	48,748	5.9	47,387
Japan	8	12,005	−6.4	11,231	−14.2	10,300	−27.6	8,696
Malaysia	3	1,744	24.6	2,173	17.6	2,050	26.8	2,211
Myanmar	2	13,807	21.5	16,776	−10.5	12,356	1.2	13,974
Philippines	3	9,459	14.1	10,797	−11.8	8,340	−4.7	9,018
South Korea	6	8,192	−13.6	7,078	−5.3	7,755	−21.9	6,401
Taiwan	7	2,798	11.8	3,128	12.8	3,156	28.0	3,583
Thailand	2	20,177	9.3	22,044	−4.7	19,230	−0.9	19,989
Total		434,136		462,472		415,129		409,743
% change				6.5		−4.4		−5.6

[1] Source: IRRI (1993).

Table 7.10. Estimated changes in total rice production predicted by the SIMRIW model for each country and in the region under the three GCM scenarios. Current actual production ('000 t) in each AEZ on a country basis are adjusted by the simulated changes in total annual production. See text and Matthews *et al.* (1995b) for explanation.

Country	AEZ	Current[1] '000 t	GFDL % change	'000 t	GISS % change	'000 t	UKMO % change	'000 t
Bangladesh	3	27,691	8.0	29,914	−10.2	24,869	−9.0	25,200
China	5	8,854	−1.7	8,700	7.3	9,501	−1.4	8,730
	6	79,872	9.7	87,596	−30.7	55,332	−17.2	66,162
	7	91,828	1.0	92,768	−10.0	82,600	−8.4	84,158
	8	2,361	−10.0	2,125	9.7	2,590	−5.8	2,223
India	1	32,807	−8.2	30,104	−19.5	26,417	−14.8	27,964
	2	49,949	9.1	54,473	10.5	55,192	−3.2	48,366
	5	227	−1.7	223	7.3	244	−1.4	224
	6	26,628	−19.9	21,339	−36.9	16,804	−92.4	2,011
	8	1,011	−10.0	910	9.7	1,109	−5.8	952
Indonesia	3	44,726	17.5	52,543	9.8	49,093	6.9	47,816
Japan	8	12,005	−10.0	10,804	9.7	13,172	−5.8	11,303
Malaysia	3	1,744	12.0	1,953	1.7	1,774	14.7	2,001
Myanmar	2	13,807	15.6	15,961	−13.8	11,896	−4.9	13,134
Philippines	3	9,459	9.4	10,345	−13.7	8,163	−5.4	8,944
South Korea	6	8,192	1.6	8,323	13.8	9,323	−8.8	7,468
Taiwan	7	2,798	2.4	2,866	5.1	2,941	20.7	3,378
Thailand	2	20,177	6.4	21,461	−11.6	17,842	−7.3	18,696
Total		434,136		452,409		388,860		378,730
% change				4.2		−10.4		−12.8

[1] Source: IRRI (1993).

Bearing these assumptions in mind, the predicted changes in overall production for each country and the whole region are shown in Tables 7.9 and 7.10 and in Figs. 7.9a – f for the two models. In this case, the ORYZA1 model predicted changes in regional rice production of +6.5%, −4.4%, and −5.6% for the GFDL, GISS, and UKMO scenarios respectively. The corresponding changes predicted by SIMRIW were +4.2%, −10.4% and −12.8%. Figure 7.10 shows a comparison of the yield changes predicted by the two crop models for each of the countries in Tables 7.9 and 7.10. The agreement is good ($P < 0.0001$), with SIMRIW tending to predict more negative changes than ORYZA1. Differences between these weighted figures and the previous average values are due to the large production of China and India in comparison with the other countries. Again the ranking of the GCM scenarios is the same for the two crop models, but the range of changes from −12.8% to +6.5% makes it clear that

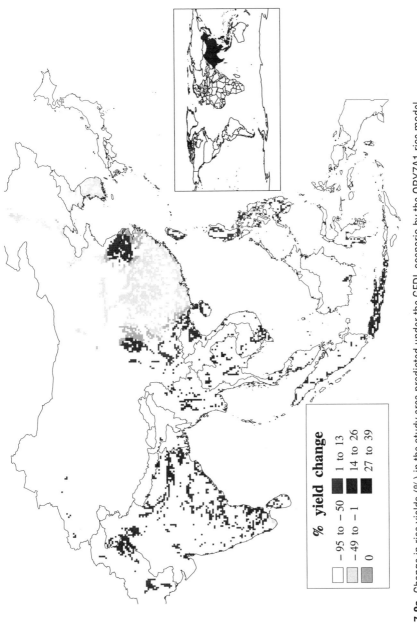

Fig. 7.9a. Change in rice yields (%) in the study area predicted under the GFDL scenario by the ORYZA1 rice model.

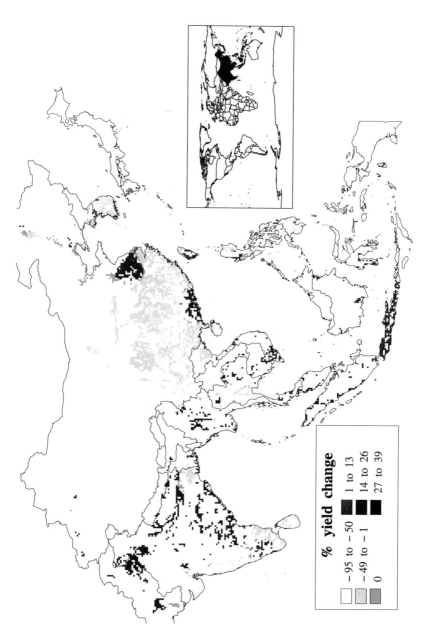

Fig. 7.9b. Change in rice yields (%) in the study area predicted under the GISS scenario by the ORYZA1 rice model.

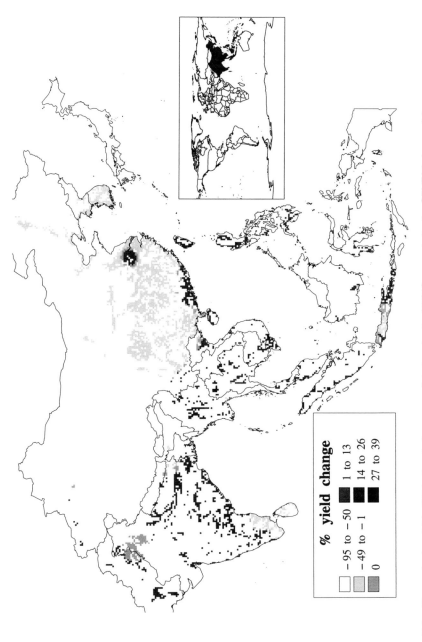

Fig. 7.9c. Change in rice yields (%) in the study area predicted under the UKMO scenario by the ORYZA1 rice model.

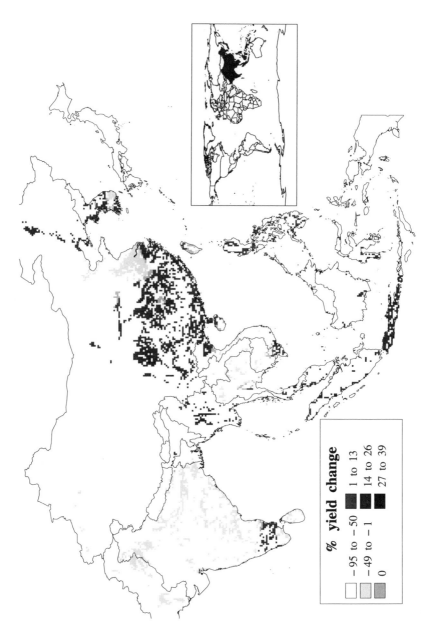

Fig. 7.9d. Change in rice yields (%) in the study area predicted under the GFDL scenario by the SIMRIW rice model.

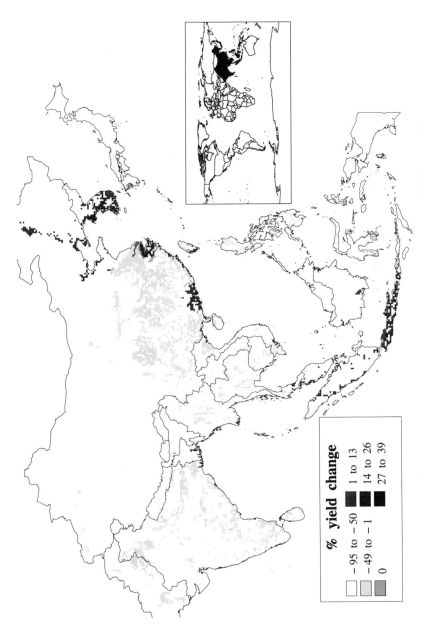

Fig. 7.9e. Change in rice yields (%) in the study area predicted under the GISS scenario by the SIMRIW rice model.

% yield change

- 95 to - 50   1 to 13
- 49 to - 1   14 to 26
0   27 to 39

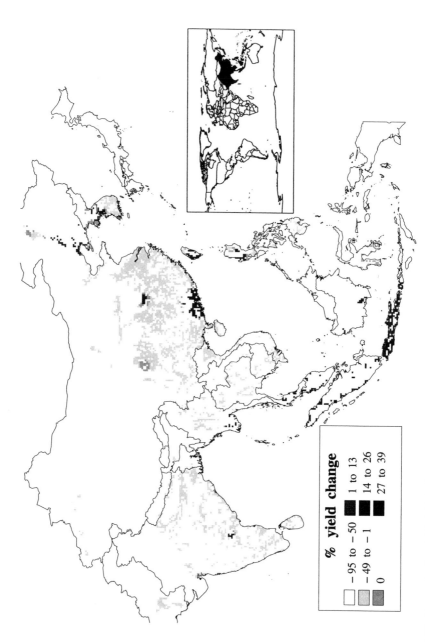

% yield change

☐ −95 to −50	■ 1 to 13	
◩ −49 to −1	■ 14 to 26	
▨ 0	■ 27 to 39	

Fig. 7.9f. Change in rice yields (%) in the study area predicted under the UKMO scenario by the SIMRIW rice model.

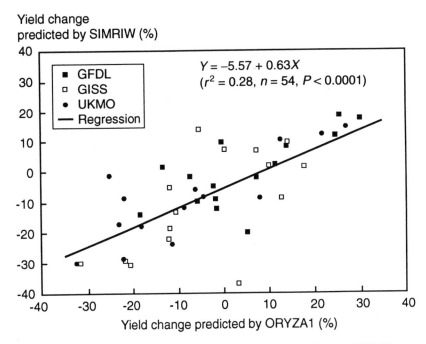

Yield change
predicted by SIMRIW (%)

$Y = -5.57 + 0.63X$
$(r^2 = 0.28, n = 54, P < 0.0001)$

- GFDL
- GISS
- UKMO
— Regression

Yield change predicted by ORYZA1 (%)

Fig. 7.10. Comparison of the yield changes predicted by the ORYZA1 and SIMRIW crop simulation models under the GFDL, GISS and UKMO scenarios. Points are the mean values predicted for each country or zone in Table 7.9.

any predictions of changes in rice production for the region depend very much on the combination of climate change scenario and the crop model used. The average across both crop models and all three GCM scenarios indicates a –3.4% decline in overall regional rice production.

Some interesting points emerge from this analysis. Averaged across all three GCM scenarios, the mean change in total production for China was predicted by ORYZA1 to be –4.2% and by SIMRIW to be –8.4%. Both of these values, particularly that of SIMRIW, compare very closely with the –7.4% predicted by Zhou (1991). Much of this decrease was due to the effect on yields in FAO-AEZ 6, which contains a large proportion of the current rice-growing areas of China. Jin et al. (1993) predicted a similar decline in irrigated rice yields in this region. Similarly, both crop models predict decreases in production for Thailand under the GISS and UKMO scenarios; similar conclusions were made in a UNEP study (UNEP Greenhouse Gas Abatement Costing Studies, 1992). Changes in production in both of these countries is likely to have serious repercussions on regional trading patterns, as China is the major importer of rice in the region (43% of total regional imported rice, Table 1.1), while Thailand is one of the major rice-exporting countries of the region (87% of total regional exported rice, Table 1.1). Similarly, in Bangladesh, the predicted decline in production by both crop models in the GISS and UKMO scenarios is likely to increase the country's need to import even more rice than at present. On the other hand, countries such as Indonesia, Malaysia, and Taiwan, and parts of India and China, are all predicted to benefit from the changes in climate predicted by the three GCMs used. There are, however, some striking discrepancies in predicted changes between the three scenarios. In FAO-AEZ 6 of China, as one example, the SIMRIW model predicts a +9.7% increase in yields under the GFDL scenario, but a −30.7% decrease in the GISS scenario. The ORYZA1 model predicts similar, but not so extreme, differences. These large fluctuations are mainly due to the sensitivity of spikelet sterility to temperatures in the region of 33°C, where a difference of 1°C can result in a modest yield increase becoming a large yield decrease (see Sections 7.5.5 and 7.6.1). In many areas, therefore, it seems that the accuracy of any prediction of changes in rice production depends on the exact nature of changes in the climate there; until there is some consensus in predictions of climate change for an area, therefore, accurate prediction of potential production changes is difficult.

7.5.5 Spikelet sterility

It is interesting to investigate the causes of the effects of elevated temperatures in more detail. The weather station at Khulna in Bangladesh was taken as an example. The simulated potential yield and its components using the standard weather data of 1984 and with the changes predicted by each of the three GCM models is shown in Table 7.11. It can be seen that the main reason for the reduced yield under the GISS scenario is the high daily maximum temperature around the time of flowering, resulting in a spikelet fertility of only 49%. Spikelet sterility is a crucial determinant of yield in rice, both at high and low temperatures, and is considered in more detail in the next section.

Table 7.11. Simulated yield and yield components at Khulna, Bangladesh, in 1984, and for the 2 × CO$_2$ scenarios of the three GCMs. Sowing date was 15 April (Aus season) and the flowering date was around mid-July. T_{max} at flowering is the maximum daily temperature on the day of flowering.

Scenario	Yield (kg ha^{-1})	Spikelet no. (m^{-2})	Filled fraction (%)	Grain no. (m^{-2})	Grain wt (mg grain^{-1})	T_{max} at flowering (°C)
Current	8,546	35,391	95.4	33,763	21.8	31.5
GFDL	9,749	38,545	95.4	36,772	22.8	32.1
GISS	4,857	37,470	48.9	18,320	22.8	36.6
UKMO	9,196	38,098	91.0	34,685	22.8	33.9

7.6 Mitigation Options

7.6.1 Varietal adaptation

The present analysis is based on the assumption that varietal characteristics are the same in the future as now. However, it is highly likely that, given the time scale involved, plant breeding programs will develop new varieties more closely adapted to the gradually changing conditions, thereby mitigating the negative, and enhancing the positive, effects of this change. Two possible adaptations that may occur are the use of varieties more tolerant to higher temperatures in the low-latitude regions, and the use of longer maturing varieties to take advantage of the longer growing season in high-latitude areas (Matthews *et al.*, 1995c).

Fraction of fertile spikelets

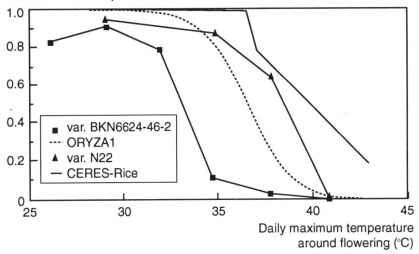

Fig. 7.11. Relationships between daily maximum temperature (°C) around flowering and spikelet fertility. CERES-Rice response calculated with $T_{min} = T_{max} -6$°C. Varietal data from Satake & Yoshida (1978).

Considerable variation between varieties in tolerance to high temperatures exists (Satake & Yoshida, 1978), although this does not seem to be related to the location of origin of the variety (Yoshida, 1981). Figure 7.11 shows the function relating spikelet fertility to daily maximum temperature for the ORYZA1 and SIMRIW models (same) and the CERES-Rice (Godwin *et al.*, 1993) model, along with data from two actual varieties. Although the ORYZA1 function is based on data from a *japonica* variety (Uchijima, 1976), it falls midway between the extremes of N22, an *indica* variety from India, and

BKN6624-46-2, an *indica* variety from Thailand. Spikelet sterility in CERES-Rice, on the other hand, is insensitive to high temperatures until 36°C is reached. This is probably the main reason for the difference in results between the present study and that of Rosenzweig *et al.* (1993), as, in the absence of spikelet sterility, yields are likely to increase as the temperature rises from 30°C to 36°C due to the lengthening of crop duration.

As an example of the possible effect of an increase in tolerance to higher temperatures, it is assumed that the sensitivity of spikelet sterility is shifted by 2°C, taking the ORYZA1 response to near that of the variety N22. The simulated effects of this adaptation under the GISS scenario are shown in Fig. 7.12 for the site in Madurai, India. Under current conditions, the temperature at the

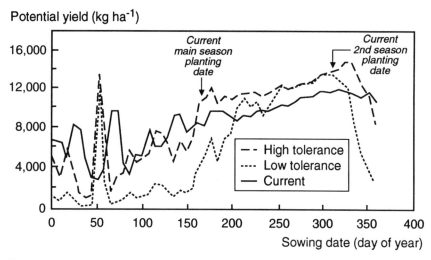

Fig. 7.12. Effect on potential yield of increasing tolerance of spikelet fertility to high temperatures, Madurai, India.

time of flowering for the main season is already high, and when the GISS scenario is imposed without the varietal adaptation, a large decrease in yield due to spikelet sterility is predicted. With the adaptation included, the yields rise to the pre-scenario level again. Thus a small shift in the response of spikelet fertility to temperature, well within the genotypic variation observed in current varieties, can offset the detrimental effect of the change in climate, even for the 'worst-case' scenario used in this example.

The effect of high temperature on spikelet sterility is difficult to model, as its effect seems to be limited only to the time of flowering, which takes place in the morning. There is some evidence that 'tolerant' varieties, in reality, avoid the high day temperatures by flowering earlier in the morning (Imaki *et al.*, 1987).

The accuracy of modeling temperature effects on spikelet sterility, therefore, depends on the diurnal time course of temperature, and on the accuracy of the model in predicting flowering date. This is particularly true where temperatures vary considerably from day to day; an error by the model in estimation of flowering date of only one day could give results that differ significantly from the observed data. This problem may be mitigated to some extent by the fact that, due to non-synchrony of tiller development, flowering in rice occurs over an extended time period of 7–10 d, so that the effect of fluctuating temperatures may be smoothed. Nevertheless, the sensitivity of spikelet sterility to temperature is a factor that must be taken into account in evaluating model predictions of the effect of climate change on rice production.

A number of researchers have suggested that the lengthening of the growing season in the northern latitudes can allow yield increases from the planting of longer maturing varieties (e.g. Yoshino *et al.*, 1988b; Horie, 1991; Okada, 1991). The effect of this possible adaptation was simulated for the Shenyang site in northern China. It was assumed that the main difference between short- and long-maturing varieties is the length of the vegetative period before panicle initiation, as it has been found for many varieties that the durations of the panicle formation and the grain-filling periods, when corrected for temperature differences, are similar regardless of total crop duration (Vergara & Chang, 1985).

The GISS scenario was used, allowing the sowing date to be advanced by 31 days from the current date of 16 May. The duration of the variety Ishikawi (77 Dd) was used as a base, and was incremented in 1 Dd intervals up to 96 Dd,

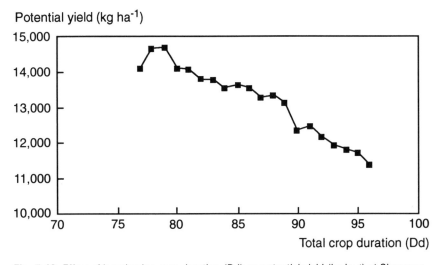

Fig. 7.13. Effect of lengthening crop duration (Dd) on potential yield (kg ha⁻¹) at Shenyang, China.

representing approximately 123 to 146 actual days in the Shenyang environment. Results are shown in Fig. 7.13. As total crop duration increased, simulated potential yields declined from around 14.5 t ha⁻¹ to 11.5 t ha⁻¹, indicating
that using longer-maturing varieties may not necessarily give extra yield. The
main reason for this was that extending the duration delayed the panicle formation and grain-filling periods until later in the season when incident solar
radiation and temperatures are lower. If positive correlations between panicle
formation duration, grain-filling duration, and total crop duration are found,
longer maturing varieties may be of benefit; a longer vegetative period is only
advantageous as long as it ensures that a closed canopy is reached by the time
of panicle initiation; any further vegetative growth after the canopy is closed
has little effect on growth of the reproductive components of the plant, and may
even represent a competitive sink for assimilate, or loss of assimilate through
unnecessary maintenance respiration.

7.6.2   Adjustment of planting dates

The analyses in Section 7.5 assumed that planting dates would not be changed
by future farmers. However, one of the consequences of a rise in temperature at
high latitudes is a lengthening of the period in which rice can be grown.
Shenyang in northern China, for example, is characterized by cold winters
from October to April, when daily mean temperatures may fall as low as −10°C,
and a cool summer period from May to September, when rice can be grown. All

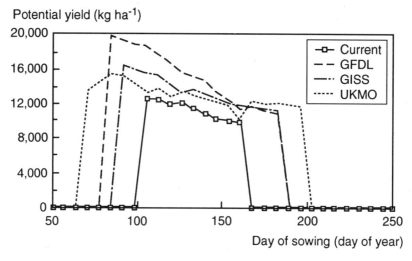

Fig. 7.14. Effect of climate change on the 'sowing window' for rice variety Ishikawi at
Shenyang, China.

of the GCMs predict an increase in both temperature and solar radiation at this location with an increase in CO_2 level. Using these predictions, the model indicated that primarily because of the higher temperatures, the period in which rice may be sown is lengthened considerably from around 70 d to 140 d in the case of the UKMO model (Fig. 7.14). If the planting date is kept the same as present (i.e. mid-May), farmers could expect an average 24% increase in potential yield with the changed climate, but if planting dates are advanced by a month, as becomes possible, yields could be increased by around 43%, due to the panicle formation and grain-filling stages receiving higher quantities of solar radiation.

On the other hand, at Madurai in southern India (latitude 8.5°N), rice yields from the main planting were significantly decreased under the predicted climate changes – an increase of 4°C from an already high temperature during the growing season resulted in a large number of sterile panicles, so that hardly any grain was produced (Fig. 7.15). The crop model indicates that by delaying planting from June until the end of July, these high temperatures could be avoided, but if this is done, it is not possible to plant another crop in the second growing season in November before high temperatures are again encountered. Overall annual production is therefore likely to decline. The advancement or delaying of planting dates must also depend on rainfall distribution, a factor not taken into account in the model at present.

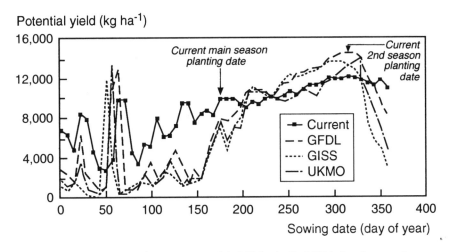

Fig. 7.15. Effect of climate change on potential yield (kg ha⁻¹) of IR64 at various sowing dates at Madurai, India.

7.7   Conclusions

As with previous studies, an increase in temperature was found to generally reduce rice yields, and an increase in CO_2 level to increase rice yields. However, the exact effect on existing crop yields at a specific site depends on the combination of these two factors at that site. The uncertainty with which the two factors, particularly temperature, can be predicted in future climate scenarios places a limit on the accuracy of estimation of the likely effect of these changes on crop yields.

Nevertheless, the use of 'fixed-change' scenarios and two crop simulation models allows some evaluation of the sensitivity of rice production in Asia to changes in CO_2 level and temperature. The ORYZA1 model predicted that overall rice production in the region would change by +6.5%, −4.4%, and −5.6% in the GFDL, GISS, and UKMO scenarios respectively, while the corresponding figures predicted by the SIMRIW model were +4.2%, −10.4%, and −12.8%. For each crop model, the GFDL scenario was the most benign and the UKMO the most severe, corresponding to the magnitude of the temperature increases predicted by each GCM. The main cause for differences in the predictions of the two crop models was the way in which leaf area development and crop growth rate was calculated, and in the routines describing phenological events in the crop.

However, for both crop models, the predicted changes varied considerably between countries. Both models predicted yield declines under the GISS and UKMO scenarios in Thailand, Bangladesh, and western India, while increases were predicted for Indonesia, Malaysia, and Taiwan, and parts of India and China. Such changes are likely to have a significant effect on future trading relationships within the south-east and eastern Asian region.

The response of spikelet sterility to temperature emerged as a major factor determining the differential predictions of the effects of climate change on each scenario. Due to the extreme sensitivity of this factor to temperatures below 20°C and above 33°C, a small difference in mean temperature can result in a positive yield change turning into a large decline, due to lower numbers of grains being formed.

Varietal adaptation, such as tolerance to high temperatures, was shown to be capable of ameliorating the detrimental effect of a temperature increase in currently high-temperature environments. For spikelet fertility, the level of adaptation required is within the range of genotypic variation currently observed. Changes in management practices, such as advanced planting dates in the higher latitudes, could give increased yields, but the use of longer-maturing varieties at such sites, as suggested by some authors, may result in lower yields, due to the grain formation and ripening periods being pushed to less favorable conditions later in the season.

7.8   Acknowledgments

The authors wish to acknowledge the help of Cecille Lopez in preparing input data, making the model simulation runs, and compiling the output data for analysis. The authors also wish to thank Zeny Pascual, Anaida Balbares, and Vic Micosa of the Climate Unit, IRRI, for their work in checking and preparing the weather data from the 68 sites used in the study.

THE IMPACT OF CLIMATE CHANGE ON RICE PRODUCTION IN SELECTED ASIAN COUNTRIES

II

Rice Production in Japan under Current and Future Climates   **8**

T. HORIE, H. NAKAGAWA, M. OHNISHI, AND J. NAKNO

Faculty of Agriculture, Kyoto University, Kyoto 606, Japan

8.1 Introduction

Japan consists of four main islands and about 4000 smaller ones, lying in a crescent shape between 20° and 45°N latitude and 123° and 146°E longitude off the eastern coast of the Eurasian continent. Japan is in FAO-AEZ 8 (see Chapter 7), characterized as temperate with summer rainfall. The climate is humid, temperate, and oceanic with four distinct seasons. The population of Japan in 1990 was over 120 million, with about 60% of this occupied in tertiary industry, 30% in secondary industry, and only 6% in the agricultural sector. Agriculture contributes only about 3% to gross domestic product.

Rice is the staple food of Japan, providing about 25% of the daily calorie intake per person; its cultivation, therefore, has an important influence on the politics, economy and culture of the country. Although 70% of its overall grain needs are imported, the country is nearly self-sufficient in rice. The area devoted to rice cultivation is controlled by the government to maintain the production at roughly ten million tons of rice per year. In recent years, however, due to a change in eating habits as standards of living have risen, rice consumption has declined, so that potential rice production now exceeds the demand. To prevent over-production of rice, policies have been promoted to encourage the alternative use of ricefields with the result that in 1990, of the 2.8 million ha of ricefields, 0.8 million ha (30%) were planted to other crops, while the share of rice in gross agricultural production fell to 2%. The distribution of rice is also partially controlled by the government, who distributed 21% of the total in 1991, with the remainder being distributed by agricultural cooperatives.

The annual pattern of temperature and solar radiation defines the rice-growing season roughly from April to October, although there is considerable variation with latitude. Rainfall during the rainy season in June and July is also necessary for successful cultivation, but excessive rainfall at the end of the rainy season can cause local flooding. Significant year-to-year variations in climate often disrupt rice production, resulting in over-production in favorable years, and fears of collapse in adverse years. The cool summer that Japan experienced in 1993, for instance, brought about a 25% reduction in national rice production, forcing the country to import approximately two million tonnes of rice.

Most ricefields are in the plains of the major river basins, although significant numbers are also found in valleys and terraces, where it may be grown at altitudes of up to 1400 m in the central region of the main island. A distinguishing characteristic of Japanese rice culture is its cultivation in the higher latitudes, with the northern limit of rice cultivation being about 44°N. To minimize the effects of low temperatures in these areas and at high altitudes, very early-maturing varieties have been developed. Management techniques such as the use of good-quality older seedlings raised under plastic covers for early transplanting, deep water irrigation to protect the crop from low night temperatures, windbreak nets, and application of organic matter to improve soil fertility have also aided production in these areas.

Almost all rice in Japan is produced in paddy fields flooded with water supplied through well-developed waterways. Thus, unlike in many other rice-producing countries, water plays a negligible role as a yield-reducing factor. The major climatic factors, therefore, that influence rice production in Japan are temperature, solar radiation, and the strong winds and heavy rainfall associated with typhoons. Although significant proportions of the total production are lost every year due to typhoon damage, the location and timing of typhoon incidence is unpredictable; temperature and solar radiation, therefore, are the main factors that produce locational and yearly variations of rice yield in Japan. Cool summer temperatures in northern Japan (Hokkaido and Tohoku districts) can cause severe yield reductions in these areas, resulting in a significant impact on the national rice production, as was the case in 1993.

Soil fertility has been gradually declining because of the increase of part-time farming, which has resulted in shallow ploughing by rotary tillers and a reliance on application of chemical fertilizers. Deep ploughing and the application of organic matter are important for improving soil fertility.

In spite of substantial farm subsidies and price support provided by the government, rice farming cannot compete with other economic activities, and income from it is lower than non-agricultural earnings. Farming operations have been fully mechanized; in 1991 for example, 98% of rice seedlings were transplanted using mechanical transplanters. Nevertheless, the production cost is many times higher than in tropical Asia due to exorbitant land prices and the high opportunity cost of farm labor. Younger people are attracted to the cities, so that rice cultivation is now carried out mainly by older people. There is also a shift in consumer preference to 'environment-friendly' rice grown without heavy use of agrochemicals; some farmers have started producing rice through organic farming methods to meet this demand.

8.2 Agroecological Zones and Current Rice Production in Japan

8.2.1 Agroecological zones

The islands of Japan are divided by a central mountain range along their axis into two major parts facing the Sea of Japan and the Pacific Ocean. This topographical feature, together with the range in latitudes described above, produce a diversity of climates. The general trend is for warm climates in the south and cool climates in the north; in addition, the climate in the area facing the Sea of Japan has dry summers and wet winters, while that in the area facing the Pacific Ocean has moist summers and dry winters. Reflecting this climatic diversity, a variety of cropping systems has also developed.

These features of climates and cropping systems were classified into 14 major agroecological zones (AEZs) by Ozawa (1962) (Fig. 8.1). The climatic conditions, possible numbers of crops per year, and major crops in each AEZ, are shown in Table 8.1. Zones I to IV are on Hokkaido island where severe winter and snow cover restricts crop production to only one crop per year. Zones V and VI are located in the northern part of Honshu island where three crops every two years are possible. Zones VII to XIII occupy central and western Honshu, and parts of Shikoku and Kyushu islands, where the climate is moderately warm and two crops per year are possible. Zone XIV is located in the southern parts of Kyushu and Shikoku islands, where a warm climate enables more than two crops per year, including double cropping of rice. Although a limit on the amount of rice cultivation is set in AEZ I and II by their cool climates, rice is the major crop in all of the other AEZs. The mountainous topography of Japan limits the amount of cultivable land, and hence rice-producing areas, mainly to the coastal parts of the islands (Fig. 8.1).

The areas of rice cultivated, yield, and total production in each AEZ are shown in Table 8.2. In general, higher yields are obtained under the cooler climates (AEZ III to VI, and IX), while the yield in the warmer areas (AEZ XI to XIV) is lower. This coincides with the model prediction that the optimum temperature for *japonica* rice is 22–23°C (see Fig. 4.11 in Chapter 4).

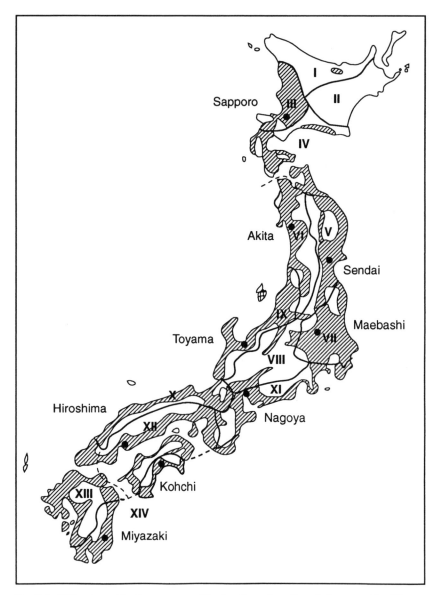

Fig. 8.1. AEZs, rice cultivation areas, and the location of weather stations used for this study. The rice cultivation area is denoted by hatching and weather stations by circles.

Table 8.1. AEZs in Japan and their characteristics (adapted from Ozawa, 1962).

Zone	Name of zone	Warmth index[1] (°C month^{-1})	Annual precip. (mm)	Frostless period (days)	Snow cover period (days)	Moisture type[2]	Crops year^{-1}	Major crops
I	Tempoku	50–55	800–1000	120–140	120–140	Dry-S/moist-W	1	Wheat, oats, potato, very limited rice
II	East-Hokkaido	45–50	c. 1000	120–140	c. 100	Moist-S/moist-W	1	Spring wheat, potato, pasture, no rice
III	Central-Hokkaido	60–70	c. 1200	140–160	c. 140	Dry-S/moist-W	1	Rice, wheat, beans, sugar beet, potato
IV	South-Hokkaido	65–75	1200–1400	160–180	70–90	Moist-S/moist-W	1	Rice, wheat, beans, maize, potato
V	East-Tohoku	75–85	c. 1200	150–170	< 50	Moist-S/moist-W	1.5	Rice, wheat, beans, maize, potato
VI	West-Tohoku	80–90	1600–2000	c. 180	100–120	Moist-S/wet-W	1.5	Rice, wheat, maize, snow limits winter crops
VII	Kanto	90–110	1300–1600	180–220	25<	Moist-S/moist-W	2	Rice, wheat, potato, sweet potato
VII	Tozan	c. 90	c. 1300	160–180	c. 40	Moist-S/moist-W	2	Rice, wheat, beans, apple, grape
IX	Hokuriku	100–110	2400–2800	c. 200	100–140	Dry-S/wet-W	2	Rice, wheat, beans, sweet potato
X	Sanin	c. 110	c. 1800	200–220	c. 50	Dry-S/wet-W	2	Rice, wheat, barley, beans, sweet potato
XI	Tokai	c. 120	c. 2000	220–240	0	Moist-S/moist-W	2	Rice, wheat, beans, vegetables, orange
XII	Setouchi	110–120	1200–1400	200–240	0	Dry-S/dry-W	2	Rice, barley, sweet potato, orange, grape
XIII	Kyushu	120–130	1600–1800	c. 220	0	Moist-S/dry-W	2	Rice, wheat, barley, beans, sweet potato
XIV	Nankai	130–140	2400–3000	240–280	0	Wet-S/moist-W	2+	Rice double cropping possible

[1] Warmth index is sum of monthly mean temperature above 5°C.
[2] Moisture type denotes pattern of annual moisture distribution. S and W stand for summer and winter, respectively.

Table 8.2.   Rice cultivation area, average yields, and production in each AEZ in Japan (calculated from Japanese Ministry of Agriculture, Forestry and Fisheries, 1991).

Zone	Cultivated area ('000 ha)	Yield* (t ha⁻¹)	Production ('000 t)
I	4	4.55	18
II	5	4.75	24
III	113	5.09	575
IV	57	5.38	307
V	202	5.10	1030
VI	254	5.74	1458
VII	294	4.64	1364
VIII	87	4.94	430
IX	237	5.19	1230
X	148	4.79	709
XI	114	4.62	527
XII	193	4.67	901
XIII	206	4.87	1003
XIV	81	4.38	355

* Yield figures are brown rice at 14% moisture.

8.2.2   Rice cropping calendar

Table 8.3 shows the main events in the rice crop calendar in the respective AEZs in Japan. Except for Hokkaido, where early spring temperatures are too cool for rice cultivation, the seeding and transplanting dates are generally earlier in the north than in the south. This general trend is also true for the heading and harvesting dates. This northwards trend of earlier transplanting is to allow flowering at the warmest period of year to avoid cool-temperature-induced sterility of spikelets during the reproductive stage. Rice cultivars raised in Hokkaido (AEZs I to IV) have practically no photoperiod sensitivity, but, towards the south, the photoperiod sensitivity of cultivars generally increases.

8.2.3   Historical rice production

Changes in the national rice yields and production over the period 1883–1987 are shown in Fig. 8.2. Owing to improvements in rice production technology, a steady increase in the yield has occurred. Total production increased in proportion to the yield level up until 1970 when a policy of restricting the areas of rice cultivation started; thereafter the relation between production and yield weakened. Despite the general trend of an increase in yields, considerably large year-to-year variation is apparent. The frequent occurrence of abnormally low yields reflect cool-summer damage (CSD) in northern Japan.

Table 8.3. Approximate dates of main phenological events in rice crop calendar in the central prefecture in each AEZ in Japan (Japanese Ministry of Agriculture, Forestry and Fisheries, 1990).

AEZ	Place	Seeding	Transplant	Heading	Harvest[1]
III/IV	Hokkaido	18 April	24 May	27 July	21 Sept
V	Miyagi	6 April	7 May	3 Aug	24 Sept
VI	Akita	17 April	18 May	6 Aug	30 Sept
VII	Gunma	19 May	18 June	24 Aug	19 Oct
VIII	Nagano	21 April	21 May	11 Aug	9 Oct
IX	Toyama	9 April	4 May	1 Aug	11 Sept
X	Tottori	1 May	25 May	17 Aug	13 Oct
XI	Aichi	1 May	26 May	19 Aug	11 Oct
XII	Hiroshima	23 April	17 May	13 Aug	4 Oct
XIII	Saga	21 May	16 June	27 Aug	12 Oct
XIV	Miyazaki	14 May	9 June	25 Aug	20 Oct

[1] Harvest time approximately coincides with maturity.

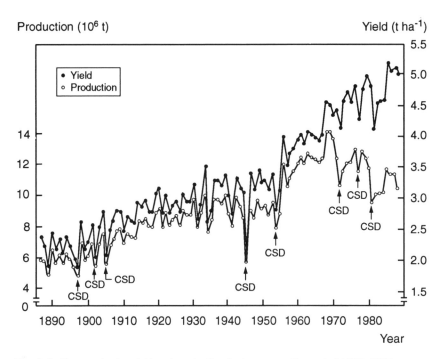

Fig. 8.2. Changes in rice yield and production in Japan over the period 1883–1987 (Japan Ministry of Agriculture, Forestry and Fisheries, 1988). CSD denotes 'cool-summer damage'.

Table 8.4. Recorded average yields of brown rice (kg ha^{-1}) for the period 1971–1990 in representative prefectures in each AEZ in Japan.

AEZ	Place	1971	1972	1973	1974	1975	1976	1977	1978	1979	1980	1981	1982	1983	1984	1985	1986	1987	1988	1989	1990
III, IV	Hokkaido	2730	5000	4790	5030	4460	3610	5040	5360	5020	3850	4130	5010	3550	5510	4970	5260	4720	5120	5260	5400
V	Miyagi	4530	5040	4850	4460	5160	4330	4680	5380	5180	3830	4340	4490	4720	5290	5480	4960	5240	3810	4840	5630
VI	Akita	4950	5160	5450	5560	5760	5140	5830	5790	5530	5470	4940	5810	5720	6130	6020	6070	5970	5450	5640	5630
VII	Gunma	3670	4280	4430	3680	4340	3740	3920	4520	3890	3680	3960	3220	3790	4510	4680	4560	4750	4010	4520	4740
VIII	Nagano	5010	5160	5450	4840	5630	4890	5700	5620	5490	4840	5370	4610	5320	5850	5680	5790	6070	5410	5540	5990
IX	Toyama	4430	4610	4720	4790	5070	4660	5040	4950	4800	4480	4940	5110	4810	5250	5190	5250	5230	4840	4790	5240
X	Tottori	3810	4470	4810	4560	4880	4520	4590	5020	4700	3640	4770	4800	4490	5280	4870	5270	4790	4920	4820	4960
XI	Aichi	3550	3150	4260	3990	4320	4180	4540	4680	4470	4230	4480	4080	4370	4640	4610	4630	4640	4640	4630	4680
XII	Hiroshima	4090	4240	4740	4580	4620	4510	4660	4850	4710	3890	4790	4490	4720	5090	4940	4950	4620	4950	4880	4690
XIII	Saga	4870	5460	5350	5270	5300	4630	5310	5160	5350	4210	5210	4960	4850	5410	4530	5400	4650	5540	5350	5160
XIV	Miyazaki	3660	3970	3930	4000	4120	3770	4260	4460	4210	4000	4270	3990	4230	4520	4510	4690	4290	4660	4380	4530

Table 8.4 shows the changes in rice yield for the period 1971–1990 in a representative prefecture in each AEZ. Although the average yield is higher in northern Japan (AEZ III to IX), the variability between years is also higher in these regions due to the occurrence of CSD. This is seen clearly in Fig. 8.3 which shows changes in the July–August temperature anomaly and rice yield in Hokkaido for the period 1881 to 1993. The July–August temperature was chosen because the reproductive development stage, when the crop is most sensitive to temperature, normally occurs in this period. Reflecting advances in rice production technology, a marked trend of yield increase with respect to time can be seen. This trend can be approximated by a third-order polynomial equation. However, despite the general trend, there is a large year-to-year variation in yield, which coincides with the variation in temperature, indicating that

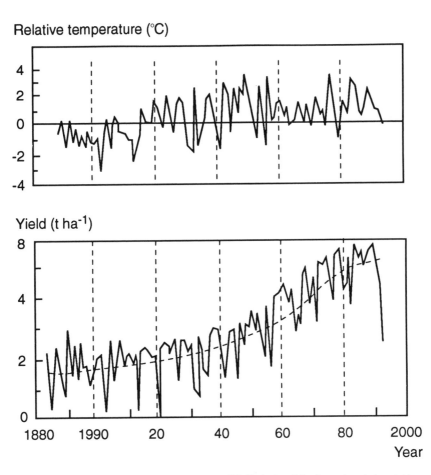

Fig. 8.3. Temperature deviations from average (20°C) during July–August and rice yield for the period 1881–1993 in Hokkaido, Japan.

summer temperature is a major cause for the yield variation in northern Japan. The yearly yield trend represented by the third-order polynomial can be considered to predict the average yield of a given year under a normal climate taking into account any technological developments. The 'yield index', defined as the ratio of the actual yield in a given year to the yield predicted by the polynomial equation for that year, was plotted against the July–August mean temperature of the same year (Fig. 8.4). Yields in Hokkaido (AEZ III and IV) and Tohoku (AEZ V and VI) drop sharply with the decrease of July–August temperature from a threshold. A similar trend can also be seen in the yield in

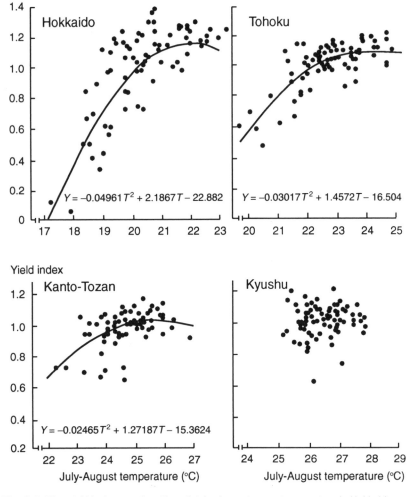

Fig. 8.4. Rice yield index as a function of July–August mean temperature in Hokkaido, Tohoku, Kanto-Tozan and Kyushu districts in Japan (NIAS, 1975).

Kanto-Tozan (AEZ VII and VIII), but with less temperature effect. In Kyushu (AEZ XIII and XIV), no clear temperature effect can be seen, where the yield variations are mainly due to differences in solar radiation, along with typhoon and pest damage.

The above analysis on regional rice yield in relationship to climate indicates that higher yields are obtained in the cooler regions compared with the warmer regions, and that yearly variation of temperature is the major cause for a large year-to-year variation of the yield. This implies that the predicted temperature changes under a changed global climate will have a significant influence on rice production in Japan. It is vitally important, therefore, to assess the impacts on regional rice yield so as to provide a basis for counter measures such as cultivar improvements and alterations of cropping seasons and cultivation technologies.

8.3   Model Parameterization and Testing

Previously, Horie (1987, 1993) has quantitatively explained much of the yield variation in Japan as functions of temperature and solar radiation using the simulation model SIMRIW (Simulation Model for Rice–Weather relationships). For the purpose of evaluating the impact of climate change on rice production, SIMRIW was modified to include processes describing: (i) the direct effect of atmospheric carbon dioxide (CO_2) on growth; and (ii) the high-temperature-induced spikelet sterility of rice (Horie, 1993). The modified model was then used for impact assessments of doubled CO_2 and climate change on rice yield in three representative rice-producing prefectures in Japan. This preliminary analysis indicated that doubled CO_2 and the associated climate change had different effects on rice yield in different locations, but the study was not sufficiently detailed to clarify the overall effect on rice production in the whole country (Horie, 1993).

The objective of the present study is to assess the effects of elevated CO_2 and climate change on rice yield in the major rice-producing areas in Japan. For this purpose, an agroecological zoning approach was taken, with the assumption that climate and cropping systems are relatively homogeneous within each zone. The two rice models, SIMRIW and ORYZA1, described in Chapters 3 and 4, were calibrated and validated for the current climate conditions in Japan, and then used to assess the likely effect on rice production under a range of climate scenarios, including those predicted by the GISS, GFDL, and UKMO General Circulation Models (GCMs; see Chapter 5).

8.3.1   Parameterization of SIMRIW and ORYZA1 for Japanese varieties and climatic conditions

Both SIMRIW and ORYZA1 models are process-oriented models for simulating growth and yield of irrigated rice; SIMRIW was developed by a rational simplification of underlying physiological processes, and hence requires only a limited number of crop parameters (Chapter 4), while ORYZA1 is a more comprehensive model in which detailed physiological processes of growth and yield formation are described (Chapter 3). Both models require as inputs daily values of solar radiation, minimum and maximum temperatures, and atmospheric CO_2 concentration, while ORYZA1 additionally requires leaf nitrogen concentration.

The *japonica* varieties Ishikari, Sasanishiki, Koshihikari, Nipponbare, and Mizuho were used to represent the range of cultivars grown in the different AEZs in Japan. Since the original parameter set installed in ORYZA1 for simulating phenological development of the *indica* variety IR72 could not explain the observed phenology of these cultivars, the phenology submodel of ORYZA1 was replaced by that of SIMRIW with cultivar specific parameters. The values of these parameters specific to each cultivar are shown in Table 8.5.

Table 8.5. Values of cultivar specific parameters for each cultivar used for the simulations.

Parameter	Variety				
	Ishikari	Sasanishiki	Koshihikari	Nipponbare	Mizuho
G_v (d)	51.0	51.0	51.3	59.6	59.9
A	0.290	0.510	0.365	0.361	0.408
T_h (°C)	17.6	17.1	17.8	18.5	17.9
B	0.660	0.530	0.566	0.558	0.607
L_c	23.6	16.0	16.0	16.2	15.7
DVI^*	1.0	0.25	0.23	0.39	0.0
h_m	0.42	0.40	0.38	0.36	0.34

Parameters G_v, A, T_h, L_c, and DVI^* characterize the phenological development for a cultivar as given in Equation 4.5, and h_m is the potential harvest index of a cultivar as given in Equation 4.13. Values of other crop parameters are common for all cultivars used.

8.3.2 Model performance

Since testing of the two models against field experimental data has already been shown in Chapters 3 and 4 of this book, this chapter focuses on testing using historic climate and yield data in different rice-producing areas in Japan. For this, reported farmers' yields for the period 1979–1990 in the five representative prefectures, Hokkaido (AEZ III and IV), Miyagi (AEZ V), Gunma (AEZ VIII), Aichi (AEZ XI), and Miyazaki (AEZ XIV) were used. Daily weather data for the corresponding period from one weather station for each prefecture were used;

these were Sapporo, Sendai, Maebashi, Nagoya, and Miyazaki for the Hokkaido, Miyagi, Gunma, Aichi, and Miyazaki prefectures, respectively. The geographical locations of these weather stations are shown in Fig. 8.1 and in Table
8.6. The varieties Ishikari, Sasanishiki, Koshihikari, Nipponbare, and Mizuho
were used for the prefectures Hokkaido, Miyagi, Gunma, Aichi and Miyazaki,
respectively. These cultivars either are, or were, the most widely grown cultivars in each prefecture.

Table 8.6.   Details of the weather stations used in the study.

Site ID	Weather station	Latitude	Longitude	Representing AEZs
1	Sapporo	43° 03′ N	141° 20′ E	I, II, III, IV
2	Akita	39° 43′ N	140° 06′ E	VI
3	Sendai	38° 16′ N	140° 54′ E	V
4	Maebashi	36° 24′ N	139° 04′ E	VII, VIII
5	Toyama	36° 42′ N	137° 12′ E	IX
6	Nagoya	35° 10′ N	136° 58′ E	XI
7	Hiroshima	34° 22′ N	132° 26′ E	XII
8	Kohchi	33° 33′ N	133° 32′ E	XIV
9	Miyazaki	31° 55′ N	131° 25′ E	XIV

Since the two models calculate the potential yield attainable under a given
climate, simulated yields by both models are much higher than the actual
yields recorded for each prefecture. We have defined a 'technological coefficient' (K) as the ratio of actual farmers' yields to simulated yields, which can
then be expressed as a function of time (see Chapter 4 for details). By assuming a
linear relationship between K and year, the simulated yield (Y_p) may be converted to farmers' yield (Y_a) using the relation

$$Y_a = [b_0 + b_1 (t\text{-}1)] Y_p, \qquad (8.1)$$

where t represents the years since 1979, and b_0 and b_1 are the coefficients of the
linear regression between K and time.

A multiple regression analysis was made between the actual yield (Y_a) and
simulated yield (Y_p) by using Equation 8.1 for each prefecture. The values of
coefficients of Equation 8.1 thus obtained are shown in Table 8.7. Similar
values of b_1 are required to convert simulated yields from both SIMRIW and
ORYZA1, with the highest value in Gunma ($b_1 = 0.022$–0.026) and the lowest
in Aichi ($b_1 = 0.004$–0.005). This suggests that the increase in rice yields due to
technological advancement is about 2.2–2.6% per year in Gunma and
0.4–0.5% per year in Aichi.

Table 8.7.   Values of regression coefficients (b_1) and (b_0) of regression equation 8.1, which is used to determine the technological coefficient as a function of year in the respective prefectures.

Prefecture	Cultivar	SIMRIW		ORYZA1	
		b_0	b_1	b_0	b_1
Hokkaido	Ishikari	0.712	0.014	0.448	0.017
Miyagi	Sasanishiki	0.695	0.010	0.529	0.015
Gunma	Koshihikari	0.651	0.026	0.513	0.022
Aichi	Nipponbare	0.675	0.004	0.575	0.005
Miyazaki	Mizuho	0.595	0.009	0.508	0.011

Using Equation 8.1 with the coefficients given in Table 8.7, the simulated yields of the two models were converted to the yields that are expected from weather conditions under the cultivation technology level of a given year for five prefectures over a 12-year period. Figure 8.5 shows the converted yield plotted against actual farmers' yields in each year in each prefecture. SIMRIW explained year-to-year variations of farmers' rice yields in the five prefectures with an $r^2 = 0.691$, and ORYZA1 with $r^2 = 0.618$. Thus, SIMRIW explained variations of regional farmers' yields in Japan slightly better than ORYZA1. Taking into account the fact that variations in rice yield are not only due to variations in temperature and solar radiation, but also to those in typhoon, pests and disease damage, both SIMRIW and ORYZA1 satisfactorily explain the regional yield variations in Japan in relation to weather conditions.

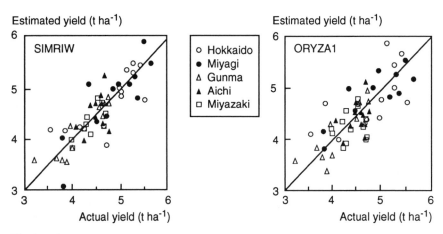

Fig. 8.5. Comparison between reported rice yields and those simulated by SIMRIW (left) and ORYZA1 (right) for five representative prefectures in 1979–1990.

8.4   Effect of Climate Change on Rice Production in Japan

8.4.1   Input data

For predicting the effects of elevated CO_2 and the resulting climate on rice yields in various AEZs in Japan, daily weather data for the twelve years from 1979 to 1990 in nine representative weather stations (Table 8.6) were selected. From this data, the 'average' climate for each location was synthesized by averaging daily weather values over the twelve years for each location. This average climate was then used as the baseline to evaluate the effects of climate change on rice yields. Five future climate scenarios were examined: 450 ppm CO_2 concentration with no change in climate, 450 ppm CO_2 with a +2°C temperature rise, and the 2 × CO_2 climates predicted by the GFDL, GISS, and UKMO GCMs respectively. The future climate conditions were created by adding the monthly temperature changes in each scenario to the current daily maximum and minimum temperatures of the same month, and by multiplying relative changes of monthly solar radiation by current daily solar radiation values. Of the climate scenarios predicted by the three GCMs under a 2 × CO_2 concentration, the temperature rise is most drastic in UKMO, moderate in GFDL and smallest in GISS. While the GFDL model predicts reductions of solar radiation in most parts of Japan in most seasons under 2 × CO_2 levels, the GISS and UKMO models predict increases, particularly in the latter.

8.4.2   Effect of fixed increments of temperature and CO_2 on potential yields

Table 8.8 shows the predicted change in rice yield under each climate scenario from that of the base (current) climate for the nine locations investigated. Overall, SIMRIW predicted that a 100 ppm increase in CO_2 concentration over the current level will increase rice yield by 7–8% at all the locations, whereas ORYZA1 predicted that yields would be increased by 16–17%. These differences are due to the different functions used to describe the response of crop growth rates to changes in CO_2 concentration (Fig. 7.3). Both estimates compare favorably, within the degree of error involved, with long-term CO_2 experimental data on rice (Imai *et al.*, 1985; Baker *et al.*, 1990a; Kim *et al.*, 1992).

Figure 8.6 shows SIMRIW predictions of the effect of an increase in 100 ppm CO_2 with +2°C temperature rise on the relative yield change from the present. Under this climatic scenario, a small negative effect is predicted only in AEZ XI (Tokai district) where the current maximum temperature is very high, but for the other AEZs, a small positive effect is predicted.

Table 8.8. Predicted change in potential rice yields from current values at various locations under different CO_2 and climate conditions.

Scenario	Sapporo	Akita	Sendai	Maebashi	Toyoma	Nagoya	Hiroshima	Kohchi	Miyazaki
SIMRIW									
450 ppm	+6.7	+8.0	+7.8	+7.7	+7.7	+7.7	+7.7	+7.7	+7.8
450+2°C	+0.7	+6.2	+3.8	+5.2	+2.4	−6.8	+1.9	+3.1	+7.3
GFDL	−2.8	+12.5	+7.3	+7.1	−10.5	−33.9	−21.1	−15.3	+3.7
GISS	+13.3	+26.0	+19.9	+22.3	+18.5	−2.2	−17.4	−29.0	+1.2
UKMO	+8.4	+24.0	+17.17	+13.6	+13.6	−40.9	−47.1	−36.7	−10.0
ORYZA1									
450 ppm	+16.7	+16.4	+16.2	+17.2	+17.4	+16.2	+16.0	+16.7	+17.0
450+2°C	−14.4	−8.9	−14.6	+0.1	−8.6	+0.1	−3.3	+1.1	+2.3
GFDL	+8.8	+6.3	+4.4	+3.5	−5.9	−1.8	+0.1	+4.3	+5.6
GISS	+25.5	+23.5	+21.7	+23.6	+25.1	+21.4	+8.1	+12.0	+7.8
UKMO	+6.9	+18.4	+13.9	+15.0	+16.5	−16.7	−23.0	−14.3	−3.7

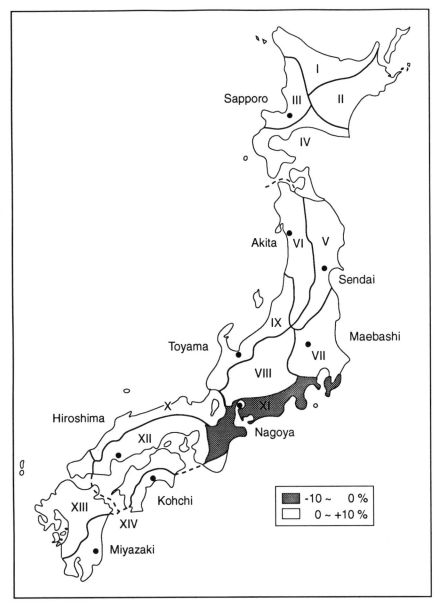

Fig. 8.6. Effect of a 100 ppm increase in CO_2 concentration and +2°C temperature rise on regional rice yield in Japan (prediction by SIMRIW).

8.4.3   Effect of predicted GCM scenarios on potential yields

Figure 8.7 shows the effects simulated by SIMRIW under the GFDL, GISS and UKMO scenarios on the relative change in yield from the present. In AEZs X and XIII where we did not have actual weather data, the relative yield changes under the respective scenarios were interpolated from those in adjacent AEZs. Although the predicted effects of a $2 \times CO_2$ climate on rice yields in Japan were quantitatively different between scenarios, the direction of the effects were similar. Under all scenarios, it was predicted that there would be moderate positive effects on rice yield in northern and north central Japan, and negative effects in south-central and south-western Japan, with the severest negative effect in AEZ XI and XII.

The positive effects of doubled CO_2 and temperature increases on rice yield in northern Japan were because the temperatures predicted by the GCMs were

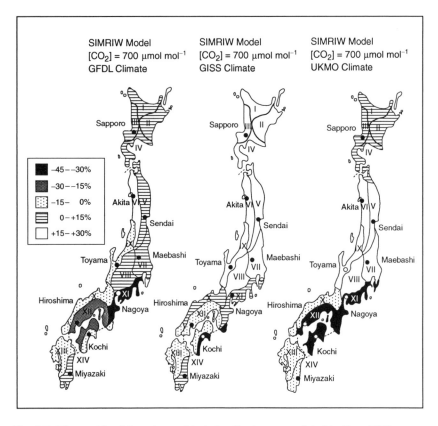

Fig. 8.7. Effects of $2 \times CO_2$ and associated climatic change predicted by three GCMs on regional rice yield in Japan (prediction by SIMRIW).

not so high as to affect spikelet fertility, and because the direct effect of doubled CO_2 more than compensated for the negative effects of warming. The yield decreases in south-central and south-western Japan were due to the predicted GCM temperatures being high enough not only to shorten rice growth duration, but also to bring about extensive panicle sterility. It is known that temperatures higher than about 35°C during the flowering period markedly reduce spikelet fertility (Satake & Yoshida, 1978; Matsui & Horie, 1992), and the results shown in Fig. 8.7 indicate that daily maximum temperatures under $2 \times CO_2$ climates frequently rise above 35°C during the flowering period in south-central (AEZ XI) and south-western Japan (AEZ XII, XIII, and XIV). Indeed, in AEZ XI, more than 30% yield reduction was predicted under the GFDL and UKMO climate scenarios.

8.4.4   Probability analysis on effects of $2 \times CO_2$ climates on rice yield

Probability analyses were made on effects of doubling CO_2 and climate change on rice yield for representative locations, by using daily weather data in the 1979–1990 period at Sapporo (AEZ I to IV), Sendai (AEZ V), Nagoya (AEZ XI) and Miyazaki (AEZ XIV) as base climates, and the climate change scenarios described previously. Figure 8.8 gives the results of the probability analysis by SIMRIW, and Fig. 8.9 those by ORYZA1, where the simulated rice yield in each of twelve years under the three different climate scenarios was plotted as cumulative probability distributions. Since the cumulative probability distributions simulated by SIMRIW and ORYZA1 were similar, further discussion is related to the results from the SIMRIW simulations.

At Sapporo in Hokkaido (AEZ I to IV), the average predicted yield under the current climate was 5.27 t ha^{-1} with a coefficient of variation (CV%) of 9.7%. Increases of +6%, +22%, and +15% were predicted for the $2 \times CO_2$ climates of the GFDL, GISS, and UKMO, respectively. The GISS scenario gave the largest yields of the three GCMs due to the smallest temperature rise and to increased solar radiation levels.

Since CSD hardly occurred under the $2 \times CO_2$ GFDL and GISS scenarios, the yield variability was also reduced. The temperature rise predicted by the UKMO scenario is so large that it causes high-temperature-induced sterility of spikelets in the warm years even as far north as Hokkaido. In general, though, a $2 \times CO_2$ climate will substantially increase the average yield and reduce the yield variability in Hokkaido.

The largest positive effect of $2 \times CO_2$ climates was predicted in Tohoku district (AEZ V and VI). At Sendai, the predicted yield increase was +14% to +26%, depending on the GCM scenario used. Predicted $2 \times CO_2$ climates reduced the yield variability in Tohoku by reducing the CV% from 15% at the present to less than 10%.

The most catastrophic effect of a $2 \times CO_2$ climate was predicted at Nagoya

Cumulative probability

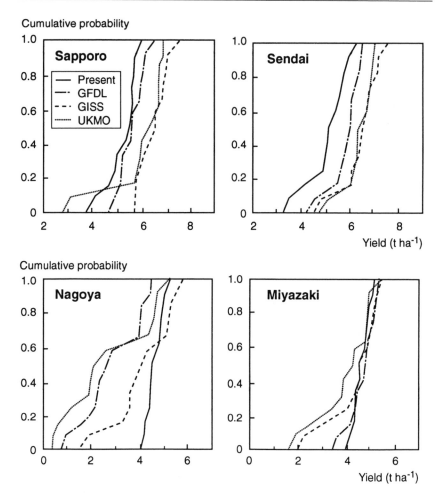

Fig. 8.8. Cumulative distribution functions for rice yield at four representative locations in Japan under three scenarios of global climate change (prediction by SIMRIW).

in Aichi (AEZ XI), where reductions in average yield of between 8% and 37% were predicted, depending on the scenario used. The yield variability was also predicted to significantly increase from the current CV% of 7% to between 27% and 61%. These catastrophic effects are due to Nagoya having the highest daily maximum temperature during current summer conditions of all the locations investigated, so that any further warming increases the possibility of high temperature-induced spikelet sterility. Similarly, at Miyazaki in Miyazaki prefecture (AEZ XIV), between 0% and 13% yield reduction is predicted for a $2 \times CO_2$ climate, depending on the scenario. Yield variability was also predicted to increase from the current 4.7% to between 11% and 26% under a $2 \times CO_2$ climate.

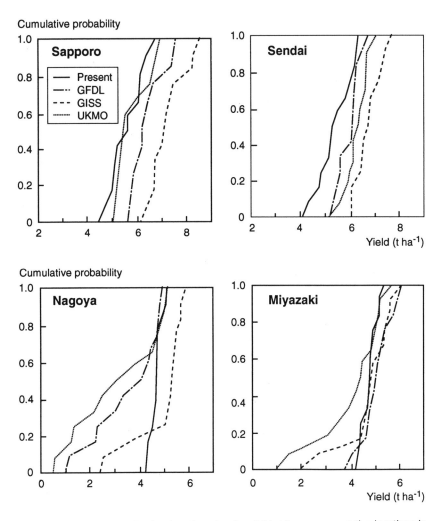

Fig. 8.9. Cumulative distribution functions for rice yield at four representative locations in Japan under three scenarios of global climate change (prediction by ORYZA1).

8.5   Conclusions

The above analysis indicates that a $2 \times CO_2$ climate will substantially increase rice yield and its stability in the northern and north-central Japan, and that, in the south-central and south-western Japan, it will decrease the yield and its stability to a marked extent. By aggregating these regional effects over the whole country, it would seem that under a changed climate the average rice yield of Japan will not change significantly from the current level. However, the variability of yields is likely to increase, reflecting increased high-temperature-induced spikelet sterility in south-central and south-western Japan.

The above analysis assumes the use of current cultivars and cropping seasons. It may be possible to reduce the predicted instability by adjusting planting dates, so as to avoid high-temperature-induced spikelet sterility by allowing the flowering period to escape the highest temperatures. This strategy, however, may reduce average yields, because the hottest season in Japan is also associated with the highest solar radiation. Grain formation and grain filling in rice has been found to be strongly influenced by solar radiation levels during the reproductive period. Further studies are required to quantitatively assess to what extent the predicted effects of a $2 \times CO_2$ climate on rice yield can be mitigated through the use of different cropping seasons and cultivars.

Rice Production in India under Current and Future Climates   9

S. Mohandass, A.A. Kareem, T.B. Ranganathan, and
S. Jeyaraman

*Tamil Nadu Rice Research Institute, Tamil Nadu Agricultural University,
Aduthurai 612 101, India*

9.1   Introduction

India is the world's second most populous nation (after China), with a population of 848 million in 1990, a growth rate of 2.0%, an infant mortality rate of 9.2% live births and a life expectancy of 59 years. Though it is still a predominantly rural country, even the urban population of 229 million exceeds the total populations of most developing nations.

The mainland of India extends from 8° to 37°N latitude and from 68° to 97°E longitude. To the north, the country is bounded by the Himalayan mountain range, and to the south, it narrows to form the Great Indian Peninsula protruding into the Indian Ocean. The Bay of Bengal is to the east of the peninsular, and the Arabian Sea is to the west. The mainland consists of four well-defined regions: (i) the great mountain zone; (ii) the Indo-Gangetic Plain; (iii) the desert region; and (iv) the Southern Peninsula. Due to its size, India spans five of the FAO agroecological zones (AEZs) (see Chapter 7): FAO-AEZ 1, the warm and semi-arid tropics; FAO-AEZ 2, the warm subhumid tropics; FAO-AEZ 5, the warm arid and semi-arid subtropics with summer rainfall; FAO-AEZ 6, the warm subhumid tropics with summer rainfall; and FAO-AEZ 8, the cool subtropics with summer rainfall.

Agriculture is the backbone of India's economy, providing direct employment to about 70% of the rural working population. It also forms the basis of many major industries, particularly the cotton textile, jute and sugar industries. In total, agriculture contributes about 31% to the gross domestic product, and about 25% of India's exports are agricultural products. Rice is the staple food for 65% of the total population in India, constituting about 43% of the total food grain production and 47% of total cereal production. Since the adoption of

modern high-yielding varieties in 1966, an increase in the area under cultivation, improved management practices, and widespread availability of credit
to farmers, an average annual increase of 2% in rice production has been
attained, resulting in the spectacular increase in national production from 51
to 110 million tonnes in 1991. Self-sufficiency in rice was reached by 1977.
The current average rice yield in India is 3 t ha^{-1}, and yields are continuing to
rise as a result of improved management. A small amount of high-quality
basmati (aromatic) rice is also exported.

India has the largest area under rice in the world, and consequently a large
diversity in rice-growing environments. Most rice is grown in FAO-AEZs 1, 2,
and 6. Of the 42 million ha of total harvested rice area, about 45% is irrigated,
33% rainfed lowland, 15% rainfed upland, and 7% flood-prone. Since the
major portion (55%) is rainfed, national production is strongly tied to the distribution of rainfall. In some of the states, erratic rainfall leads to drought during
the vegetative period, but later in the season the crop may be damaged by submergence due to high rainfall. In the eastern states, damage due to flash floods
can be high. Soil fertility is often another constraint; extreme soil acidity is a
problem in southern and eastern India, whereas in northern India soil salinity
and alkalinity can limit yields. Nitrogen, phosphorus, and zinc deficiencies are
also widespread. Nearly all of the rainfed area suffers from a lack of infrastructure. Moreover, many farmers are not able to afford the inputs necessary for
maximum production.

9.2   Agroecological Zones and Current Rice Production in India

9.2.1   Agroecological zones

The Planning Commission of India (Government of India, 1989) has defined
five major zones on the basis of topography, namely; (i) the Himalayas and
associated hills; (ii) the Northern Plains including the Indo-Gangetic Plains;
(iii) the Peninsular Plateau and Hills; (iv) the East Coast Plains; and (v) the West
Coast Plains. These five zones are further subdivided into a total of fifteen AEZs,
based on physical conditions, topography, soil geological formation, rainfall
pattern, cropping systems, and development of irrigation and mineral resources at the district level. Two of these were in the Himalayan region, five in
the Northern Plains, four in the Peninsular Plateau and Hills, one in the East
Coast Plain, and two in the West Coast Plain, with the fifteenth region covering
the Andaman, Nicobar, and Lakhadweep islands. Descriptions of these fifteen
zones and rice growing areas are given in Fig. 9.1 and Table 9.1. The fifteen
zones are further classified into a total of 73 subzones, taking into account soil,
geographic information, temperature, rainfall, water conditions (including
quality of water and aquifer conditions), cropping patterns, and farming
systems.

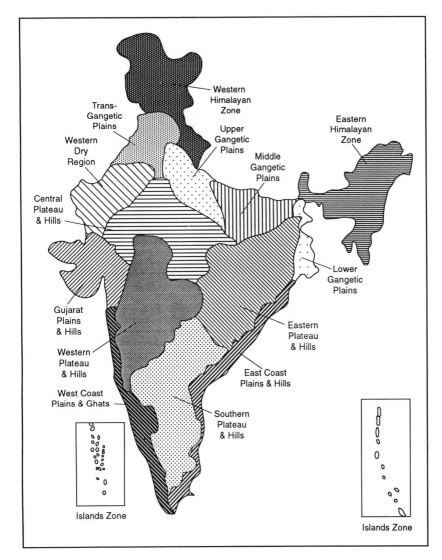

Fig. 9.1. The 15 main agroclimatic zones in India. (Source: Khanna, 1989.)

Rice cultivation is found in the West Coast region, Tamil Nadu, the coastal parts of Andhra Pradesh, the south Central region, and the Eastern, Northern, and Gujarat regions. In general, rainfall plays the most influential role in determining areas of rice production in India. Rao & Das (1971) have described in detail the relationship between rice yields and weather patterns, a summary of which is as follows:

Table 9.1. Description of the 15 agroecological zones in India, including the major crops grown in each region.

No.	Zone	Annual rainfall (mm)	Climate	Representative soil	Major crops
1	Western Himalayas	165–2000	Cold arid to humid	Brown hill, Alluvial meadow, Skeletal	W, M, R, P
2	Eastern Himalayas	1840–3528	Perhumid to humid	Alluvial, Red, loamy, Red sandy, Brown hill	R, M, J, R, M
3	Lower Gangetic plains	1302–1607	Moist subhumid to dry subhumid	Red and Yellow deltaic alluvium, Red loamy	R, J, W, R, M
4	Middle Gangetic plains	1211–1470	Moist subhumid to dry subhumid	Alluvial, Loamy alluvial	R, M, W
5	Upper Gangetic Plains	721–979	Dry subhumid to semi-arid	Alluvial	W, R, M, T
6	Trans-Gangetic plains	360–890	Extreme arid to semi-arid	Alluvial	W, M, R, S
7	Eastern Plateau and hills	1271–1436	Moist subhumid to dry subhumid	Red sandy, Red, Yellow	R, W, M, Ra
8	Central Plateau and hills	490–1570	Semi-arid to dry subhumid	Mixed Red and Black, Red and Yellow, Medium Black, alluvium	W, G, J, R, B
9	Western Plateau and hills	602–1040	Semi-arid	Medium Black, Deep Black	J, B, C, W
10	Southern Plateau and hills	677–1001	Semi-arid	Medium Black, Deep Black, Red sandy, Red loamy	J, R, Ra, Gr, Ra, J, B
11	East Coast plains and hills	780–1287	Semi-arid to dry subhumid	Deltaic alluvium, Red loamy, Coastal, alluvium	R, Gr, Ra, J, B
12	West Coast plains and hills	2226–3640	Dry subhumid to perhumid	Laterite, Red loamy coastal alluvium	R, Ra, G, To
13	Gujurat plains and hills	340–1793	Arid to dry subhumid	Deep Black, Coastal alluvium, Medium Black	R, Ga, C, B, W
14	Western dry	395	Arid to extremely arid	Desert, Gray, Brown	B, GW, Ra
15	Islands	1500–3086	Humid	Gray, Brown	Co

[1]B = bajra; C = cotton; G = gram; Gr = groundnut; J = jowar; M = maize; P = potato; R = rice; Ra = finger millet (ragi); S = sugar-cane; T = tur; To = tapioca; W = wheat; Co = coconut.

1. *West Coast region (Kerala, Coastal Karnataka, and Konkan).* The heavy south-west monsoon rain lasts for more than five months in Kerala but less than three months in Konkan. In Kerala, where three crops are grown, the first rice season begins on 1 April. The rainy days in the middle of April lead to better germination and growth in the nursery. In coastal Karnataka with only two crops, the season starts in May. The single crop in Konkan is planted in June.

2. *Tamil Nadu.* In all three seasons, rice in Tamil Nadu is grown as an irrigated crop only and 78% of the rice is planted to a first crop.

3. *Coastal Andhra Pradesh.* The first crop in Andhra Pradesh is more important in the northern districts such as Guntur. Regression analysis indicates that the weather does not markedly influence rice yields in this subdivision.

4. *South Central region.* This subdivision includes the regions of South and North (interior) Karnataka, Rayalaseema and Telengana of Andhra Pradesh, and Madhya Maharashtra and Marathwada of Maharashtra. In North and South (interior) Karnataka rainfall is an important meteorological factor in rice production. The maximum temperature in the north Karnataka subdivision and the mean daily minimum temperature during the harvest in Rayalaseema are also significant meteorological factors. In Madhya Maharashtra, the number of rainy days in July and the mean daily relative humidity at pollination and fertilization are the most important weather factors determining the yield.

5. *Eastern region.* This subdivision consists of areas of Assam (North and South), sub-Himalayan West Bengal, and Gangetic West Bengal. Rainfall in Assam is an important factor in rice production. High maximum temperatures during grain formation (6–12 Oct) have a deleterious effect on rice in South Assam, but the high range of mean temperatures are greatly beneficial to rice in North Assam. Rainfall is the dominant variable determining the yield in West Bengal. Lack of rain in the plateau or high range of mean temperatures from 23 to 29 August in the plains coinciding with early flowering depresses rice yield in Bihar. The predominant weather factor in Uttar Pradesh is the drought period from August to mid-September. The greater range of mean temperatures during the flowering depresses grain yield.

6. *Northern region (Himachal Pradesh, Punjab and Haryana).* Punjab and Haryana have similar weather factors, rainfall alone playing a part in determining rice yield. Occasional drought from early August to mid-September is of significance. Himachal Pradesh has a totally different set of agrometeorological factors, except the presowing rainfall.

7. *Gujarat region.* This region lies in the northwest of the Konkan coast, where cloudiness during the flowering phase assumes greater significance. Cloudy weather favors pollination and fertilization.

Vamadevan & Murty (1976) reported that the traditional rice-growing areas in India are exposed to less than 450 hours of bright sunshine from July to September. The west coast and major parts of eastern and Northern India receive less than 300 hours sunshine during the period. Productivity potential is closely related to the solar radiation during the crop season.

9.2.2 Rice cropping calendar

Rice is sown and harvested almost all year round in India. The largest proportion is monsoon-dependent, and is grown during the wet season (June–November) in the traditional rice-growing areas, although the availability of irrigation water allows more than one crop per year to be grown in some areas. Figure 9.2 shows the sowing and harvesting periods in each of the states (adapted from the Directorate of Economics and Statistics 1972, as quoted by Sreenivasan, 1980). Eight states grow rice only in one season; five states

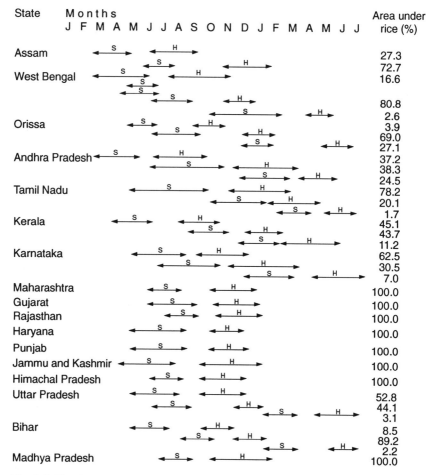

Fig. 9.2. The rice cropping calendar for each of the states of India, showing sowing and harvesting dates for each of the major rice-growing seasons. Figures indicate the percentages of the total annual sown area planted in each season (after Sreenivasan, 1980).

cultivate 69% or more area in one season. Only 7% and 3% of Karnataka and Uttar Pradesh respectively are planted to summer rice. Kerala has about the same acreage under first and second rice, and 11% under summer rice. Rice is evenly distributed between the first two seasons in Kerala, mainly because rainfall is spread over more than 150 days and transplanting of rice is practiced. In Andhra Pradesh, 96% of the rice area is irrigated and three crops are widely grown. The state accounts for 8.7% of the all-India rice area and 12% of India's total rice production.

9.2.3   Historical rice production

The historical trend of rice production in India from the period 1950 to 1989 is shown in Fig. 9.3. Although the area sown to rice only increased by 37% over the 40-year period, national production increased by 360%. This was due primarily to a general increase in yields through the use of modern high yielding varieties and improved management practices. Table 9.2 shows the areas sown, and the average yield and total production of rough rice by state in India. The most important rice-producing states are West Bengal, Andhra Pradesh, and Uttar Pradesh, which together contribute about 42% of India's total rice production.

Table 9.2.   Areas sown, average yield and production of rough rice by state in India, 1989.

State	Area ('000 ha)	Yield (t ha^{-1})	Production ('000 t)	Proportion of total production (%)
West Bengal	5,614	2.92	16,385	15.1
Andhra Pradesh	4,191	3.62	15,192	14.0
Uttar Pradesh	5,358	2.62	14,035	12.9
Punjab	1,908	5.24	9,996	9.2
Bihar	5,328	1.81	9,660	8.9
Orissa	4,392	2.17	9,521	8.8
Tamil Nadu	2,015	4.64	9,347	8.6
Madhya Pradesh	5,036	1.39	7,009	6.5
Assam	2,435	1.72	4,192	3.9
Karnataka	1,183	3.01	3,564	3.3
Maharashtra	1,520	2.29	3,478	3.2
Haryana	621	4.10	2,547	2.3
Kerala	572	2.76	1,577	1.5
Gurajat	601	2.04	1,226	1.1
Jammu & Kashmir	264	3.29	869	0.8
Total	41,038	43.62	108,598	100.0

Note: some columns do not total exactly due to rounding-up of numbers.

Area ('000000 ha)   Average yield
or production ('000000 t)   (t ha⁻¹)

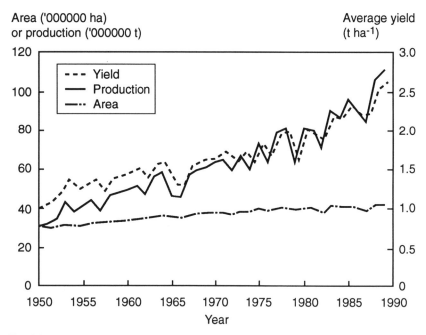

Fig. 9.3. Historical rice production statistics in India, showing the area sown, average national yields and annual national production. (Source: IRRI, 1991.)

9.3   Model Parameterization and Testing

9.3.1   Parameterization of ORYZA1 for Indian varieties and climatic conditions

The rice crop growth model ORYZA1, described in Chapter 3, was used in the present study to evaluate the effect of climate changes on rice production in India. Although different cultivars are grown in different locations and at different times of the year throughout India, it was decided for the sake of consistency to parameterize the model only for the high-yielding cultivar IR36, which is grown widely throughout the country. Values of the parameters were based on the results of a multilocation trial conducted at the Tamil Nadu Rice Research Institute (TNRRI), Aduthurai, Tamil Nadu, during the dry season (June–September) of 1993, and are shown in Table 9.3.

9.3.2   Model performance

The model was tested using data from a potential production experiment conducted at TNRRI, Aduthurai, using the five cultivars Jaya, ADT-36, IR50, IR64, and IR36. Adequate nitrogen fertilizer was applied, and care was taken

Table 9.3. Genotype coefficients and function tables (expressed as a function of developmental stage, DVS) for the variety IR36 used in the simulations. FLVTB and FSTTB are the fractions of new assimilate partitioned to the leaves and stem respectively; SLATB is the specific leaf area (cm g^{-1}), and NFLVTB is the leaf nitrogen concentration (g N m^{-2} leaf).

Genotype coefficient	Name	Units	IR36
Initial relative leaf area growth rate	RGRL	(°Cd)$^{-1}$	0.006
Stem reserves fraction	FSTR	–	0.4
Basic vegetative period duration	JUDD	Dd	20.9
Photoperiod-sensitive phase duration	PIDD	Dd	15.0
Minimum optimum photoperiod	MOPP	h	11.3
Photoperiod sensitivity	PPSE	Dd h^{-1}	0.0
Duration of panicle formation phase	REDD	Dd	26.7
Duration of grain-filling phase	GFDD	Dd	28.7
Maximum grain weight	WGRMX	mg grain^{-1}	25.4
Spikelet growth factor	SPGF	sp g^{-1}	64.9

FLVTB = 0.0, 0.56, 0.5, 0.56, 1.0, 0.0, 2.1, 0.0
FSTTB = 0.0, 0.44, 0.5, 0.44, 0.7, 0.65, 1.0, 0.4, 1.2, 0.2, 2.1, 0.0
SLATB = 0.0, 500, 0.273, 500, 0.494, 350, 0.610, 320, 0.766, 210, 0.883, 300, 1.313, 270, 1.625, 270, 2.1, 220
NFLVTB = 0.0, 0.805, 0.273, 0.805, 0.494, 1.555, 0.61, 1.25, 0.766, 1.552, 0.883, 1.42, 1.313, 1.39, 1.625, 0.955, 2.1, 0.811

to keep the crop free from pests, diseases, and weeds. The other five cultivars are similar to IR36 in morphology and growth duration. There was good agreement between observed and simulated yields for all five cultivars (Fig. 9.4). Simulated yields at other locations, using the respective weather and sowing/transplanting dates, also agreed well with observed yields.

9.4 Effect of Climate Change on Rice Production in India

9.4.1 Input data

The ORYZA1 model, calibrated for the variety IR36 as described above, was used to simulate potential rice yields under various scenarios of changed climate, including 'fixed-scenario' changes of CO_2 only (1.5 and 2 times the current level), temperature only (+1°C, +2°C, and +4°C above current temperatures), combinations of these changes, and under scenarios predicted by the GFDL, GISS, and UKMO General Circulation Models (GCMs) described in Chapter 5. In all, 15 different scenarios were simulated, details of which are given in Table 7.5. Nine locations representing the major rice-growing regions of India and for which daily weather data were available were selected, details of which are shown in Table 9.4 and Table 7.3. For each site, dates of sowing and transplanting were determined according to the appropriate cropping calendar in Fig. 9.2.

The recorded daily values of minimum and maximum temperatures were adjusted by the changes used for each scenario, and in the case of the three GCM scenarios, solar radiation was also adjusted by the predicted factor. The

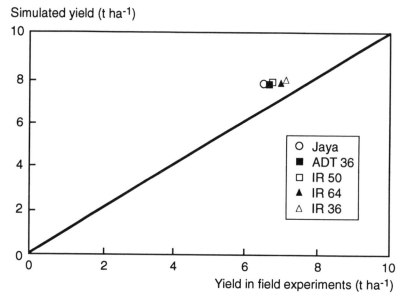

Fig. 9.4. Comparison of observed and simulated yields for different rice varieties.

Table 9.4. Details of the weather data sites used for the simulations.

Station name	Code	State	AEZ	Years available
Aduthurai	ADT	Tamil Nadu	11	1968–1992
Bijapur	BIJ	Karnataka	10	1972–1980
Coimbatore	CBE	Tamil Nadu	10	1961–1970, 1980–1981
Cuttack	CUT	Orissa	11	1983–1987, 1990
Hyderabad	HYD	Andhra Pradesh	10	1984
Kapurthala	KPR	Punjab	6	1984, 1985
Pattambi	PTB	Kerala	10	1983–1985
Madurai	MDU	Tamil Nadu	11	1986–1990
Patancheru	PAT	Andhra Pradesh	10	1978–1984

current CO_2 concentration level was assumed to be 340 ppm, so that the $1.5 \times CO_2$ scenario represented 510 ppm and the $2 \times CO_2$ scenario represented 680 ppm.

9.4.2 Current potential production

The average potential yields for the nine sites, along with their coefficients of variation (CV%), are shown in Table 9.5. Simulated yields for the main season ranged from 7.1 t ha^{-1} in Cuttack to 10.5 t ha^{-1} at Kapurthala. The higher yield in Kapurthala was due to the lower temperatures and higher solar radiation prevalent during the growing season. For the second season, yields ranged from 5.3 t ha^{-1} in Bijapur to 10.3 t ha^{-1} in Coimbatore. At most sites, with the exception of Bijapur and Patancheru, yields were higher in the second season than in the main season, but the variability was often also higher. The mean potential production of rice across all sites and years was 8.56 t ha^{-1} and 8.68 t ha^{-1} for the main and second seasons respectively, falling well within the range of 6–10 t ha^{-1} of potential rice grain yield in previous simulations for India (Penning de Vries, 1993).

Table 9.5. Mean simulated potential yields and their coefficients of variation (CV%) for each weather station. Values are averages of simulated yields for each year available for each station. Only part of one year of data was available for Hyderabad.

	Main season		Second season	
Location	Mean potential yield (kg ha^{-1})	CV%	Mean potential yield (kg ha^{-1})	CV%
Aduthurai	8,307	4.4	9,817	5.3
Bijapur	8,121	6.1	5,342	37.1
Coimbatore	9,023	8.1	10,277	2.9
Cuttack	7,129	12.6	8,958	13.2
Hyderabad	9,007	–	–	–
Kapurthala	10,503	6.4	–	–
Pattambi	7,736	0.0	8,541	9.5
Madurai	8,823	6.1	9,513	10.8
Patancheru	8,408	5.3	8,365	24.4

9.4.3 Effect of fixed increments of temperature and CO_2 on potential yields

The effects of varying the CO_2 level and temperature by fixed increments is shown in Table 9.6. Values are the average percent changes in potential yields of all the years and all the stations. When the temperature was not varied, increasing the CO_2 concentration increased the yield in the main season by an average of +28.8% and +45.2% for the 1.5 × CO_2 and 2 × CO_2 scenarios, respectively. The corresponding figures for the second season were very similar, at +28.5% and +44.6%. These values are higher than the 30% for a doubling of CO_2 suggested by Kimball (1983), but similar to those found for other countries in this report, and within the range reported by Baker *et al.* (1990a).

Table 9.6. Yield changes (%) predicted for the main and second planting seasons under fixed scenarios of increased CO_2 level and temperature increments of +1°C, +2°C, and +4°C. Values are overall means of the nine weather stations and all years available for each.

	Main season				Second season			
CO_2 level	+0°C	+1°C	+2°C	+4°C	+0°C	+1°C	+2°C	+4°C
340 ppm	0	−4.6	−8.6	−12.7	0	−17.0	−33.8	−63.6
1.5 × CO_2	28.8	22.7	17.7	11.9	28.5	6.8	−14.7	−53.2
2 × CO_2	45.2	38.3	32.6	25.9	44.6	20.1	−4.0	−47.4

Increasing temperatures decreased the yield in both seasons, but the effect was much more drastic in the second season than in the main season. When CO_2 level was not varied, the mean decrease in yield in the main season was 4% per 1°C increase, but was 16% °C^{-1} in the second season. This difference reflects the higher temperatures normally encountered at most sites in the second season, so that any further increase in temperature is likely to be to a level where spikelet sterility from high temperature damage is a major factor limiting yields.

A doubling of CO_2 level was more than able to compensate for the detrimental effect of increased temperatures up to +4°C in the main season, but was not able to in the second season. It seems clear, therefore, that yields in the second season will be much more affected by an increase in CO_2 level and the associated temperature increase than will those in the main season.

9.4.4 Effect of predicted GCM scenarios on potential yields

The predicted changes in yield for both planting seasons under the three GCM scenarios are shown in Table 9.7. In the main season, substantial yield increases, ranging from +8.8% to +40.2%, were predicted for each of the nine stations in all three scenarios. Averaged across sites, yield increases of +25.5%, +28.2%, and +28.4% were predicted for the GFDL, GISS, and UKMO scenarios, respectively. In contrast, in the second season, large decreases in yield were

Table 9.7. Changes in yield (%) predicted by the ORYZA1 rice model under each of the three GCM scenarios in the main and second planting seasons. Values are the mean of the percentage change predicted for each year available for each weather station.

Location	Main season			Second season		
	GFDL	GISS	UKMO	GFDL	GISS	UKMO
Aduthurai	40.2	28.3	24.6	42.1	10.3	−62.8
Bijapur	8.8	29.1	34.4	−87.7	−95.1	−96.0
Coimbatore	37.8	32.5	29.2	−2.3	−47.0	−78.4
Cuttack	28.8	36.9	27.5	−46.8	−62.7	16.8
Hyderabad[1]	22.1	34.0	38.5	−	−	−
Kapurthala[2]	29.6	27.8	24.2	−	−	−
Pattambi	31.7	26.7	28.9	34.6	32.9	−65.5
Madurai	18.6	15.2	19.5	−2.7	−61.3	−27.6
Patancheru	11.7	23.5	28.9	−63.4	−80.3	−96.0
Mean	25.5	28.2	28.4	−18.0	−43.3	−58.5

[1] Weather data for the second season not available.
[2] No second season crop normally grown.

generally predicted, with the corresponding mean yield changes of −18.0%, −43.3%, and −58.5% for each of the three scenarios. Yields at Patancheru and Bijapur were particularly affected; under the UKMO scenario, almost total loss of the crop (−96% decrease) occurred in both cases. At other sites, there were marked differences between the scenarios; at Aduthurai, for example, predicted changes ranged from −63% to +42%. This variation reflects the extreme sensitivity of the model, and of the rice crop, to temperature changes in the region of 32°C to 34°C, where spikelet sterility from high temperature is a major factor limiting yields.

It is interesting to investigate in more detail the underlying causes for these differences between the two seasons. For this, Patancheru was taken as an example. Daily weather data for 1980 was used as the baseline, and the effect of each GCM in the two planting seasons on the components of yield was simulated. The results of these simulations are shown in Table 9.8. In the main season, temperatures are not high enough in any of the scenarios to affect spikelet fertility, and the higher numbers of spikelets formed due to the enhanced growth of the crop between panicle initiation and flowering from the increased CO_2 level are also associated with higher grain numbers, translating directly into yield. Higher solar radiation (S_p) during the panicle formation period predicted by the UKMO scenario further enhanced spikelet number formation. In all scenarios, the duration of the crop is shortened by 4–5 days by the increased temperatures, but this is more than compensated for by the effect of the increased CO_2.

Table 9.8. Components of yield for simulated crops growing in the main and second planting seasons at Pantcheru under the three GCM scenarios. T_p and S_p are the mean daily values of temperature and solar radiation respectively over the period from panicle initiation to flowering.

Season/ scenario	Yield (kg ha^{-1})	Duration (days)	Spikelet no. (m^{-2})	Filled fraction	Grain no. (m^{-2})	T_p (°C)	S_p (MJ m^{-2} d^{-1})
Main season							
Current	8,806	126	46,573	0.954	44,431	25.2	13.8
GFDL	9,483	122	53,403	0.954	50,946	26.9	12.5
GISS	10,420	122	57,503	0.954	54,858	30.2	15.2
UKMO	10,813	121	61,143	0.954	58,330	28.7	16.2
Second season							
Current	10,854	148	68,137	0.664	45,262	26.5	23.0
GFDL	9,442	151	90,714	0.352	31,968	29.1	22.9
GISS	5,103	149	91,651	0.189	17,276	30.0	23.1
UKMO	356	158	83,330	0.014	1,206	34.5	23.7

In the second season, again many more spikelets are formed, as a result both of supra-optimal temperatures increasing the crop duration, and of the effect of the increased CO_2 level on crop growth rate. However, because of the high temperatures around the time of flowering, severe spikelet sterility occurred, with the result that grain numbers were drastically reduced. In the case of the UKMO scenario, the most severe, where daily mean temperatures during the panicle formation period were 34.5°C, more than 98% of all spikelets were sterile.

An attempt was made to estimate the effect of these predicted changes on the overall national annual rice production of India. To take into account the wide range of rice growing areas in the country, and because state-level rather than agroecological-level production figures were available, it was decided to stratify national production according to the various states, using the data presented in Table 9.2. The limited number of weather stations for which daily weather data was available also restricted this analysis, and it was necessary to group several states together to be represented by a single station. In this way, six groups were defined, which are shown in Table 9.9. Geographic proximity and similar agroecological zoning were generally the criteria for grouping, but in the case of Gujarat, it was assumed that the similar latitude and coastal environment would be best represented by Cuttack. For each group, an estimate of the proportions of rice planted in the main and second seasons was made, based on the figures given in Fig. 9.2, weighted by the annual state total production where several states were combined. Where there were more than

Table 9.9.   Stratification of states for estimation of the effects of climate change scenarios on overall national production of rice in India. Production is the combined production for each state in 1989 (from Table 9.2). Proportion of rice in the main season is taken from Fig. 9.2, weighted according to the total production from each state.

Group	Representative weather stations	States covered	Production ('000 t)	% of rice in main season
1	Hyderabad, Patancheru	Andhra Pradesh	15,192	93.0
2	Bijapur	Karnataka, Maharashtra	7,042	96.5
3	Pattambi	Kerala	1,577	88.8
4	Cuttack	Orissa, Assam, Bihar, Madhya Pradesh, West Bengal, Gurajat	47,993	85.2
5	Kapurthala	Punjab, Jammu & Kashmir, Uttar Pradesh, Haryana	27,447	100.0
6	Aduthurai, Coimbatore, Madurai	Tamil Nadu	9,347	78.2

two planting seasons (e.g. West Bengal), the second and subsequent seasons were combined into one. The resulting estimates are shown for the main season in Table 9.9. Consistently in all states, by far the largest proportion of rice is planted in the main season, ranging from 85% to 100% in all groups except Tamil Nadu with 78%. Errors in the estimates of these proportions are, therefore, not likely to be large.

In this way, the predicted change in production under a given scenario for each group was calculated by partitioning the total group production into the production for each season, adjusting these values by the predicted percent changes (Table 9.7), and summing across seasons. Total national production was then calculated by summing across groups, and the overall predicted change by comparing this value with the current annual national production. Results of this analysis are shown in Table 9.10. In all three GCM scenarios, annual national production is predicted to increase; the values are remarkably consistent at +20.1%, +22.9%, and +23.8% for the GFDL, GISS, and UKMO scenarios, respectively.

9.5   Conclusions

The results of this study indicate that the annual national rice production of India is likely to increase as CO_2 levels and temperatures increase in future climates. The source of this increase is main season crops where the fertilizing effect of the increased CO_2 level is more than able to compensate the crop for any detrimental effects of increased temperatures. These results agree with those obtained by Achanta (1993) using the CERES-Rice model (Godwin et al., 1993)

Table 9.10. Estimation of the effect of the three climate change scenarios on overall annual rice production in India. Current annual production for each stratified group is adjusted by the predicted changes in the main and second planting seasons weighted by the proportion of the groups' production from each season.

Group	Current ('000 t)	% main season	GFDL ('000 t)	% change	GISS ('000 t)	% change	UKMO ('000 t)	% change
1	15,192	93.0	16,904	11.3	18,400	21.1	18,940	24.7
2	7,042	96.5	7,420	5.4	8,782	24.7	9,138	29.8
3	1,577	88.8	2,082	32.0	2,010	27.4	1,866	18.3
4	47,993	85.2	56,447	17.6	58,654	22.2	60,446	25.9
5	27,447	100.0	35,585	29.6	35,083	27.8	34,082	24.2
6	9,347	78.2	11,951	27.9	10,532	12.7	9,987	6.9
Total	108,598		130,389		133,461		134,459	
% change				20.1		22.9		23.8

Predicted production (P_p) in each scenario is calculated as:

$$P_p = P_c [f_1 (1+C_1/100) + (1-f_1) (1+C_2/100)],$$

where P_c is current production, f_1 is the fraction of rice produced in the main season, C_1 is the predicted change (%) in the main season, and C_2 is the predicted change (%) in the second season.

using weather data for Pantnagar. While large decreases were predicted for second season crops at many of the locations in the present study due to supra-optimal temperatures being encountered, the relatively low proportion of total rice produced in this season meant that its overall effect on national production was small.

However, the predicted shift in production from the second season to the main season may mean that extra planning is necessary. Storage and milling facilities should be improved to deal with the extra main season production and to carry over some of the production for the rest of the year. Rural infrastructure in the form of roads may also need to be improved to deal with the extra transport required in moving rice from the areas of production to the points of consumption.

The large decreases in second season yields, however, may be able to be offset to some extent by selection for more temperature tolerant varieties. Considerable genotypic variation is known to exist in the relationship between daily maximum temperature and the fraction of fertile spikelets (Satake & Yoshida, 1978).

A major limitation of the present study is the few sites for which daily weather data was available. This meant that large areas of India were, of necessity, represented by a single weather station. In view of the large number of AEZs that have been identified throughout the country, it is likely that there are some errors introduced into the analysis. Future studies of the effects of climate change on rice production in India should, therefore, include a larger number of weather stations to better reflect this heterogeneity in rice-growing areas.

Rice Production in Malaysia under Current and Future Climates

10

S. SINGH, A. RAJAN, Y.B. IBRAHIM, AND
W.H. WAN SULAIMAN

Universiti Pertanian Malaysia, 43400 UPM Serdang, Selangor, Malaysia

10.1 Introduction

Malaysia in the Malay Archipelago is made up of the Malay Peninsula, Sabah, and Sarawak, located between 1° and 7°N latitude and 100° and 120°E longitude. Peninsular Malaysia, extending from the Thailand border in the north to Singapore in the south, has a coastline of some 1930 km, while Sarawak and Sabah on the northern part of Kalimantan have a coastline of about 2253 km. The population is approximately 18 million, with nearly 50% residing in the rapidly expanding urban areas. Agriculture contributes about 16%, manufacturing 29%, and services 44%, to gross domestic product.

Malaysia is in FAO-AEZ 3 (see Chapter 7), characterized as the warm humid tropics. The climate is governed by the north-east monsoon, extending from October until March, and the south-west monsoon, from May until September, although intermonsoon periods are also marked by heavy rainfall. The north-east monsoon from the South China Sea is the wettest season, occasionally causing widespread floods on the east coast of Peninsular Malaysia, particularly from November to February. The south-west monsoon period is drier, particularly for the west coast of the Peninsula, which is sheltered by the landmass of Sumatra. In general, Sabah and Sarawak experience greater rainfall than does Peninsular Malaysia. Temperature averages a constant high of 26°C with a diurnal range of 7°C, but the annual range in daily mean temperature is only 1°C.

Rice is the staple food of Malaysia, although annual per capita consumption has dropped from 100 to 79 kg over the past 20 years with changing trends in consumer preference. Thus, even though the population growth rate

exceeds 2% annually, the national requirement for rice has increased less drastically. Malaysia is about 73% self-sufficient in rice, the remainder of which is
imported mainly from its trading partners. Recently, there has been an increase
in demand for high-quality, fragrant rice. Although rice production is heavily
subsidized through input supplies (seeds, fertilizers, chemicals), price supports,
and on-farm infrastructure, over the past five years there has been a net
decrease of more than 5% in the area cultivated, most of which is simply abandoned and left idle, rather than being used for other short-term food crops. A
shortage of affordable farm labor is widely believed to be the main cause of this
decline. Although yields are predicted to increase by about 1% per year, the
decline in area is expected to continue at the rate of about 0.6% per year (Sixth
Malaysia Plan 1991–1995), but with the government policy of confining rice
cultivation to major granary areas with irrigation and drainage infrastruc-
tures, this area may decline even more rapidly.

Table 10.1.   Area under padi cultivation, annual production and imports of rice into
Malaysia for the period 1990–1991.

Year	Area under padi[1] ('000 ha)	Production ('000 t)	Imports ('000 t)	Self-sufficiency (%)
1990	650	1173	329	78
1991	661	1256	430	74
1992	666	1269	462	73

[1] Double cropped areas counted twice.

Table 10.1 shows the total rice area, annual production and imports of rice
of Malaysia for the years 1990, 1991, and 1992. Peninsular Malaysia has
about 480,000 ha of cultivated rice area, two-thirds of which is irrigated and
double-cropped, Sabah has about 50,000 ha, of which only a small fraction is
fully irrigated, while most of the 132,000 ha in Sarawak are rainfed and
upland. Because of its irrigation facilities, Peninsular Malaysia accounts for
almost 88% of total production, despite containing only 67% of the total rice
area. More than half of the total rice area is planted to modern varieties, with
indica varieties being the most common. In the irrigated and partially irrigated
areas, crop establishment is mainly by direct (broadcast) seeding although
manual transplanting is also practiced in a small percentage of this area; in the
rainfed lowlands, mainly transplanting and in the uplands, drill seeding pre-
dominates. Machine harvesting is a common practice in most irrigated aeas of
Peninsular Malaysia and is being promoted in other areas to offset the labor
shortage.

The main overall production constraint is the shrinking of the cultivated rice area, which overrides any advantage that might be expected from increasing the yield level. Moreover, in some areas, there is strong evidence of a gradual yield decline. Other constraints include periodic drought, irregular rainfall, and seasonal monsoon floods. Shortage of irrigation water in some years is also a problem. In the upland areas, ineffective terracing, and inefficient, gravity-fed irrigation systems can limit yields. In other areas, soil nutrient imbalances and low cation exchange capacities are a problem. Most farmers opt for blanket applications of chemical fertilizers rather than using organic materials.

10.2   Agroecological Zones and Current Rice Production in Malaysia

10.2.1   Agroecological zones

Figure 10.1 shows the agroecological zones (AEZs) and major rice-growing areas of Peninsular Malaysia. The focus of this study was in three regions: namely, Telok Chengai representing the Muda agricultural region; Kemubu representing the Kemubu agricultural region; and Tanjung Karang representing the Barat Laut Selangor region.

The Muda agricultural region is in the North-west agroclimatic zone and is characterized by a clear and regular dry season during the period from December to March. A secondary dry season occurs around July/August only in the extreme north and in the south of this region. Soils are mainly of marine and riverine alluvia with medium to heavy texture. Muda is the major rice-growing area of the country, and rice is the predominant crop in the region. The Kemubu agricultural region falls in the East Coast agroclimatic zone, and is characterized by a season of very heavy rainfall from October to December caused by the north-east monsoon, which often results in flooding over extensive areas in the coastal region during this period. This is followed by a regular dry season during the remaining months (January to April) of the north-east monsoon. Rice is mainly grown on the river terraces of the flood plain, and soils are of medium to heavy texture. The Barat Laut region is in the West Coast agroclimatic zone, the main climatic feature of which is the more even seasonal distribution of rainfall. Dry periods occur frequently in February and also in June/July, but these are generally very short, lasting for less than one month. The region experiences occasional strong gusts of wind and also morning rainfall which prevail during the south-west monsoon season (May to September). Soils of this region are mainly marine and riverine alluvia with medium to heavy texture. Rice is the main crop along the coastal plains in this region. All three regions described above have irrigation and drainage facilities for double

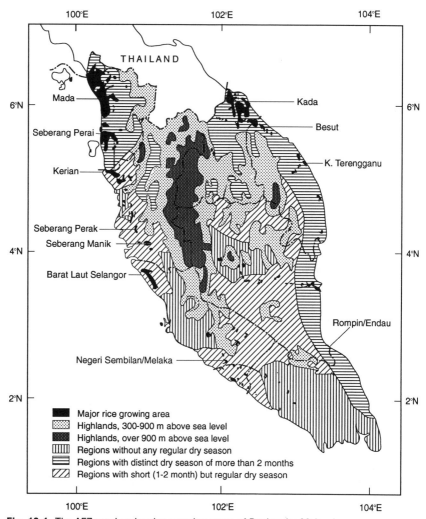

Fig. 10.1. The AEZs and major rice-growing areas of Peninsular Malaysia.

cropping of rice, and form part of the overall national agricultural development program in the region, where rice is the central crop, but with other related activities being encouraged to enhance the livelihood and income of rice farmers.

10.2.2   Rice cropping calendar

Figure 10.2 shows the cropping calendar by AEZ for rice in Peninsular Malaysia. There are two main rice cropping seasons: the main season with planting

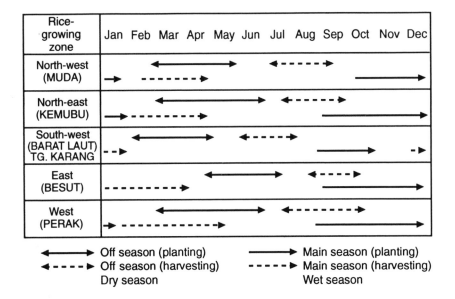

Fig. 10.2. Cropping calendar by AEZ for rice in Peninsular Malaysia.

from September to December, and the off season with planting from February through May, although these times may vary a little depending on the region. The bulk of the harvesting is done in December–February and in September for the two seasons, respectively. The main season corresponds with the wetter north-east monsoon, and the off season with the dry period before the onset of the south-west monsoon.

For the simulations, sowing dates of 9 August for the main season and 10 March for the off season were assumed. Although, in reality, there may be some slight variance around these dates, these were used for all of the three main rice-growing areas in Malaysia. Transplanting in both seasons was assumed to be at 25 days after sowing.

10.2.3 Historical rice production

Table 10.2 shows the area and historical yield data of rice from the three major rice growing areas of Peninsular Malaysia, namely, the Muda, Kemubu, and Barat Laut regions. The area sown to rice in these regions is approximately the same in both seasons. The large fluctuations in planted area between years are due to lack of irrigation facilities during the season concerned or construction activities to improve irrigation facilities during that season. In the Muda region, off season yields were lower, whereas in the Kemubu region, off season yields

Table 10.2. Historical yield data and area planted for the three major rice-growing areas of Peninsular Malaysia.

Year	Season	Telok Chengai Area (ha)	Telok Chengai Yield (kg ha⁻¹)	Kemubu Area (ha)	Kemubu Yield (kg ha⁻¹)	Tanjung Karang Area (ha)	Tanjung Karang Yield (kg ha⁻¹)
1985/86	Main	95,714	3,995	18,091	3,350	17,741	3,122
	Off	90,966	3,506	17,795	3,602	11,708	3,059
1986/87	Main	95,994	4,396	13,061	3,113	11,461	3,618
	Off	88,956	2,663	20,930	2,844	15,806	3,667
1987/88	Main	96,121	4,042	19,950	2,930	17,767	3,759
	Off	92,220	3,182	18,804	3,536	17,811	3,793
1988/89	Main	97,022	3,999	6,283[1]	2,376	17,891	3,504
	Off	92,447	3,490	23,499	3,621	17,016	4,026
1989/90	Main	97,105	4,039	21,056	3,445	17,714	4,189
	Off	92,590	3,593	24,804	3,612	8,783[2]	3,772

Main season = Wet season; Off season = Dry season.
[1] 13,787 ha not planted due to infrastructure development. Total area = 20,070 ha.
[2] 8,317 ha not planted due to infrastructure development. Total area = 17,100 ha.

were higher, and for the Barat Laut region, only small differences were observed between seasons. These can be attributed mainly to the differences in solar radiation from the time of flowering of the rice crop for a particular season at the three sites. In terms of overall total production, 54% is produced during the main season and 46% during the off season from these regions (Table 10.2).

10.3   Model Parameterization and Testing

10.3.1   Parameterization of ORYZA1 for Malaysian varieties and climatic conditions

The ORYZA1 rice potential production model (version 1.21), described in Chapter 3 was used to simulate the effect of temperature and CO_2 on growth and yield of rice. The model was calibrated for the local variety MR84 using parameters obtained from experiments at Tanjung Karang in 1988 and 1989. Table 10.3 shows the genotype coefficients and function tables used for this variety.

Table 10.3. Genotype coefficients and function tables (expressed as a function of developmental stage, DVS) for the variety MR84 used in the simulations. FSHTB, FLVTB, and FSTTB are the fractions of new assimilate partitioned to the shoot, leaves, and stem respectively; SLATB is the specific leaf area (cm g^{-1}), and NFLVTB is the leaf nitrogen concentration (g N m^{-2} leaf).

Genotype coefficient	Name	Units	MR84
Initial relative leaf area growth rate	RGRL	(°Cd)$^{-1}$	0.005
Stem reserves fraction	FSTR	–	0.2
Basic vegetative period duration	JUDD	Dd	31.6
Photoperiod sensitive phase duration	PIDD	Dd	15.0
Minimum optimum photoperiod	MOPP	h	11.3
Photoperiod sensitivity	PPSE	Dd h^{-1}	0.17
Duration of panicle formation phase	REDD	Dd	22.7
Duration of grain-filling phase	GFDD	Dd	25.0
Maximum grain weight	WGRMX	mg grain^{-1}	24.0
Spikelet growth factor	SPGF	sp g^{-1}	50.0

Main season
FLVTB = 0.0, 0.52, 0.26, 0.51, 0.50, 0.485, 0.7, 0.35, 0.95, 0.05, 1.0, 0.0, 2.1, 0.0
FSTTB = 0.0, 0.48, 0.26, 0.49, 0.50, 0.515, 0.7, 0.60, 0.95, 0.15, 1.0, 0.05, 1.1, 0.0, 2.1, 0.0
SLATB = 0.000, 470, 0.152, 470, 0.336, 330, 0.653, 280, 0.787, 210, 1.011, 190, 1.431, 170, 2.100, 170
NFLVTB = 221, 0.84, 246, 0.84, 259, 2.02, 269, 1.630, 280, 1.31, 301, 1.17, 311, 1.02, 322, 0.85, 332, 0.6, 346, 0.58, 350, 0.56

Off season
FLVTB, FSTTB, SLATB same as for the main season
NFLVTB = 69.0, 0.86, 94.0, 0.86, 107.0, 2.10, 117.0, 1.69, 130.0, 1.53, 150.0, 1.36, 160.0, 1.25, 171.0, 1.08, 181.0, 0.88, 193.0, 0.69, 206.0, 0.60, 216.0, 0.58

10.3.2 Model performance

The model was validated with experimental data for the cultivar MR84 grown in the 1990 main season, with 120 kg N ha^{-1} applied. Figure 10.3 shows the simulated and observed values of leaf area index (LAI), shoot biomass, and panicle weights. Agreement is good, particularly with LAI and biomass. For panicle weight, the model predicted higher values earlier in the season, and slightly lower values than observed at maturity.

The model was also used to predict the average regional yields for each of the three main rice-growing areas in Malaysia (Table 10.6). The model predicted yields on average 2.4 times that of the regional average yields, so the predicted yields were multiplied by a factor similar to the 'technology factor' used by Horie (Chapter 8), using a value of $1/2.4 = 0.42$. The adjusted

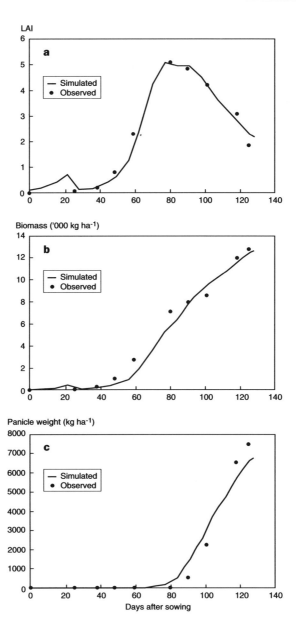

Fig. 10.3. Comparison of crop growth simulated by the ORYZA1 model using the crop parameters shown in Table 10.3 with observed values from an experiment at Tanjung Karang, 1990, using the variety MR84 with 120 kg N ha^{-1}. (a) LAI (m^{-2} m^{-2}), (b) total above-ground biomass (kg ha^{-1}), and (c) panicle dry weight (kg ha^{-1}).

predicted yields are compared with the observed regional average yields for both main and off seasons in several years in Fig. 10.4. There is a sizable variation in the observed values (2.3–4.3 t ha^{-1}) that is not explained by the model, which gives a range of only 3–4 t ha^{-1}. As the model only takes into account solar radiation and temperature, it would seem that other factors than these influence year to year variation in regional yields. Fertilizer supply, irrigation water availability, and pests and diseases would be obvious candidates. The 'technology factor' of 0.42 also suggests that management factors are limiting yield in some way, and that there is scope for considerable improvement.

10.4   Effect of Climate Change on Rice Production in Malaysia

10.4.1   Input data

The model, calibrated for the variety MR84 as described above, was used to simulate potential rice yields under various scenarios of changed climate,

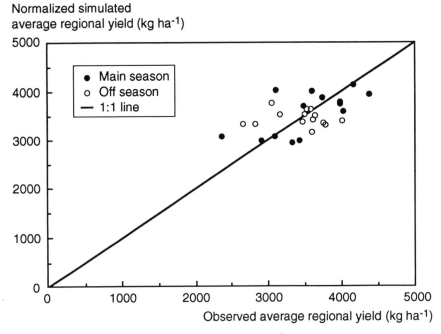

Fig. 10.4. Comparison of yields simulated by the ORYZA1 model, using the crop parameters shown in Table 10.3, with regional rice yields for the main and off seasons (Table 10.2). Observed yields were on average 0.42 of the simulated yields; the latter are therefore adjusted by this factor in a way similar to the technology factor used by Horie (Chapter 8).

including 'fixed-scenario' changes of CO_2 only (1.5 and 2 times the current level), temperature only (+1°C, +2°C and +4°C above current temperatures), combinations of these changes, and under scenarios predicted by the GFDL, GISS, and UKMO General Circulation Models (GCMs) described in Chapter 5. In all, 15 different scenarios were simulated, details of which are given in Table 7.5. Daily weather data for three weather stations (Kemubu, Telok Chengai, and Tanjung Karang) were used, details of which are given in Chapter 2 of this book. For Kemubu, eight years of data (1984–1991) were available, and for Telok Chengai and Tanjung Karang, ten years of data (1982–1991) were available.

The average predicted increments in temperature over the growing season for the three GCMs for the three regions and for the main and off seasons are given in Table 10.4. For Kemubu and Telok Chengai, the highest temperature increases were predicted by the GISS scenario, particularly during the off season. For Tanjung Karang, the UKMO scenario predicted the highest temperature increases of 4.7 to 5.3 °C. For the GISS scenario, predictions for all three sites were the same, reflecting the larger grid size of this GCM (Table 5.1).

Table 10.4. Temperature increments (°C) predicted by the three GCM scenarios for the two growing seasons in the three main rice-growing regions of Malaysia. Values are the averages over the months of each growing season.

Location	Main season			Off season		
	GFDL	GISS	UKMO	GFDL	GISS	UKMO
Kemubu	+2.41	+3.75	+3.58	+2.50	+3.76	+2.28
T. Chengai	+2.36	+3.75	+3.58	+2.38	+3.76	+2.28
T. Karang	+2.44	+3.75	+4.68	+2.20	+3.76	+5.27

Table 10.5. Yield changes (%) predicted for the main and off seasons under 'fixed' scenarios of increased CO_2 level and temperature increments of +1°C, +2°C, and +4°C. Values are overall means of the three weather stations and all years available for each.

CO_2 level	Main season				Off season			
	+0°C	+1°C	+2°C	+4°C	+0°C	+1°C	+2°C	+4°C
340 ppm	0.0	−6.4	−10.8	−19.2	0.0	−4.0	−7.6	−29.1
1.5 × CO_2	+26.6	+18.7	+13.4	+2.2	+26.8	+21.9	+16.4	−12.7
2 × CO_2	+41.7	+32.9	+27.1	+14.2	+41.8	+36.3	+29.7	−3.6

10.4.2   Effect of fixed increments of temperature and CO_2 on potential yields

Table 10.5 shows the predicted changes in yield under the fixed-change scenarios. At all CO_2 levels, and in both seasons, there was a decrease in yield as temperature increased. The decline when temperature increased by +4°C in the off season was particularly severe. Averaged over both seasons at the current CO_2 level, the predicted decline was 5.4% per 1°C increase in temperature, somewhat lower than the 7–8% °C^{-1} decrease measured experimentally by Baker *et al.* (1992b). At all temperature increments, increasing the CO_2 level increased yields. When temperature was not changed, doubling the CO_2 level increased yields by 42%, higher than the 30% suggested by Kimball (1983). Baker *et al.* (1990a) reported increased rice grain yields by as much as 50% with a doubling of CO_2 from 330 to 660 ppm, but in a later study obtained no grain yield increase due to CO_2 enrichment, which they attributed to lower levels of solar radiation (Baker *et al.*, 1992a). In our study, a doubling of CO_2 was more than able to offset the detrimental effect of a rise in temperature in all cases except the +4°C increment in the off season (Table 10.5).

10.4.3   Effect of predicted GCM scenarios on potential yields

The predicted changes in yield under the three GCM scenarios are shown in Table 10.6 for both planting seasons, and for the whole country. In the main season, substantial yield increases, ranging from +12% to +30%, were predicted for each of the three regions in all three scenarios. Averaged across scenarios, +12.4%, +22.9%, and +26.9% were predicted for Kemubu, Telok Chengai, and Tanjung Karang, respectively. At each site, the GISS scenario gave lower increases than the other two scenarios. In the off season, similar increases in yield were predicted, with the exception of Tanjung Karang under the UKMO scenario, in which a substantial decrease in yield was simulated, corresponding with the high temperature increment predicted by this scenario (Table 10.4). Mean yield changes for each of the three regions were +13.2%, +35.1%, and +7.2% for Kemubu, Telok Chengai, and Tanjung Karang, respectively.

Changes in overall national rice production were calculated by adjusting the seasonal production in each region by the predicted changes (Table 10.6), using the 1989/90 values from Table 10.2 as a baseline. These were then summed and compared with the total current production figures. In this analysis, a +25.7%, +18.7%, and +20.7% increase in production was predicted for the main planting season under the GFDL, GISS, and UKMO scenarios, respectively (Table 10.6a). Corresponding figures in the off season were +25.8%, +20.0%, and +36.4% (Table 10.6b). The predicted production for each season were then summed to give the change in annual rice production for the whole country. In this way, a +25.7%, +19.3%, and +28.6% increase was calculated

Table 10.6. Predicted effect of climate change under the three 2 × CO_2 GCM scenarios on national rice production in Malaysia, stratified according to the three main rice-growing regions. Production figures (tonnes, t) in 1989/1990 (Table 10.2) are used as the current baseline, and are adjusted by the predicted fractional changes for each region. Predicted changes are the means of all years for each weather station. Overall national production is the sum of the productions in the main and off seasons.

(a) Main season

Location	Current (t)	GFDL		GISS		UKMO	
		(% change)	(t)	(% change)	(t)	(% change)	(t)
Kemubu	72,538	14.3	82,918	10.6	80,198	12.4	81,533
T. Chengai	392,207	27.2	498,798	19.7	469,304	21.7	477,274
T. Karang	74,204	29.0	95,723	21.6	90,232	30.0	96,465
Total	538,949		677,439		639,734		650,272
Overall change (%)			+25.7		+18.7		+20.7

(b) Off season

Location	Current (t)	GFDL		GISS		UKMO	
		(% change)	(t)	(% change)	(t)	(% change)	(t)
Kemubu	89,592	8.2	96,896	1.9	91,294	29.5	116,035
T. Chengai	332,676	29.9	432,136	24.9	415,573	50.6	501,170
T. Karang	64,501	29.5	83,528	19.5	77,078	−27.3	46,892
Total	486,769		612,560		583,945		664,097
Overall change (%)			+25.8		+20.0		+36.4

(c) Overall annual national production

Location	Current (t)	GFDL (t)	GISS (t)	UKMO (t)
Kemubu	162,130	179,814	171,492	197,568
T. Chengai	724,883	930,934	884,877	978,444
T. Karang	138,705	179,251	167,310	143,357
Total	1,025,718	1,289,999	1,223,679	1,319,369
Overall change (%)		+25.7	+19.3	+28.6

for the GFDL, GISS, and UKMO scenarios, respectively (Table 10.6c). It thus seems that under all the envisaged scenarios of climate change in the next century, rice production in Malaysia will increase significantly above current levels.

Table 10.7. Simulated components of yield for rice grown in the main season at Kemubu. Weather data for 1989 is used for the current scenario, which is then adjusted by the increments of temperature and solar radiation predicted by the GFDL, GISS, and UKMO GCM scenarios. Flowering temperature is the mean daily temperature around the time of flowering.

Scenario	Yield (kg ha⁻¹)	Duration (days)	Spikelet no. (m⁻²)	Grain filling (%)	Grain no. (m⁻²)	Grain size (mg grain⁻¹)	Flowering temp. (°C)
Current	7,202	129	31,730	95.4	30,270	20.5	26.3
GFDL	8,324	125	36,936	95.4	35,237	20.3	28.6
GISS	8,520	128	37,874	95.4	36,132	20.3	29.9
UKMO	8,656	127	36,921	95.4	35,222	21.1	29.8

It is interesting to look more closely at the factors causing these predicted yield increases. Table 10.7 shows the components of yield of simulations run for current conditions and under the three GCM scenarios. Weather data for 1989 at Kemubu is used as the baseline, which is then adjusted by the predicted changes in solar radiation and temperature for each GCM. The yield increase (ranging from +15.6% to +20.2%) is explained by an increase in spikelet number and hence grain number. The crop duration is shortened slightly in all scenarios, but this is more than offset by the fertilizing effect of CO_2 enhancing the crop growth rate. This results in more spikelets being formed over the period between panicle formation and flowering. Despite increments of 2 to 3°C under the scenarios, temperatures at the time of flowering are not high enough to cause spikelet sterility (Table 10.7) so that the sink size is significantly increased, which is then capable of being filled by the increased assimilate production during the grain-filling period. Satake & Yoshida (1978) reported that spikelet sterility was induced by high temperatures in excess of 35°C almost exclusively on the day of anthesis.

Crop duration was found to be only slightly affected during the main season at all three locations, but was hastened by as much as 8 to 11 days during the off season under the UKMO scenario. Baker et al. (1992a) also reported a hastening of the maturation period of their rice crop by as much as ten days under different temperature regimes. The harvest index was also reduced with increasing temperatures.

10.5   Conclusions

The results in this study indicate that, in general, under the three GCM scenarios considered, national rice production in Malaysia is likely to increase significantly. This is a result of temperatures during the growing season in the current climate being about 26°C, which, even with the temperature increments predicted by the GCMs, do not rise to a level where spikelet fertility is likely to be influenced by high temperatures. This is in contrast to many other locations in this study. The increased temperatures also do not shorten the duration of the crop sufficiently to negate the beneficial effect of the increased CO_2 level, contrary to the findings of another study (UNEP Greenhouse Gas Abatement Costing Studies, 1992) which predicted a 12–22% yield decline under the GISS scenario.

However, the validity of these findings rests on the assumptions made in the simulations. While it seems unlikely that temperatures will increase by more than that which the GCMs predict, there is some uncertainty in the actual CO_2 level associated with the scenario. We have assumed that the $2 \times CO_2$ scenario represents twice the current level of 340 ppm; in reality, the predicted scenarios are for an 'equivalent' doubling of CO_2, which, taking other greenhouse gases into consideration, means that the actual concentration of CO_2 would be less than 680 ppm. Actual yield increases may, therefore, be less than we have predicted in this study. We have also not included the effect of the changed climate on water availability. Increasing temperature greatly increases water use, so that more severe water stress may be experienced by rice crops in the future, particularly in the off season. Another study (UNEP Greenhouse Gas Abatement Costing Studies, 1992) predicted an increase of 15% in irrigation demand, which would affect small-scale farmers more.

In order to meet these challenges, counteractive steps must be undertaken that should include, among others:

1. *Breeding for new cultivars.* Varieties that are tolerant to higher temperatures that are likely to be encountered under the changed climatic scenario possibly through genetic engineering. Selection of cultivars that are more efficient in their water use which are adapted to changed climatic conditions. Varieties which have a longer ripening phase to ensure adequate grain filling would be desirable.

2. *Water management.* Water storage facilities should be enlarged and efficiently managed. Efficient delivery and distribution of water as and when needed to the units in orderly sequence is essential.

3. *Management practices.* Land leveling must be done for more efficient water distribution and crop management. Agronomic practices (e.g. fertilizer application, weed control, pest and disease management) may need to be adjusted under the changed climate. It might also be worthwhile to consolidate small farm holdings into larger units to take advantage of the economy of scale to improve management efficiency.

4. *Cropping systems.* Adjustments in the dates of planting may be necessary for the efficient utilization of water and energy resources under the changed

climatic scenario. Conservation of moisture between cropping seasons especially for the dry season is proposed by seeding green covers that can also build up soil fertility.

With adequate attention to the measures proposed above, it is hoped that rice production in the tropics will not be adversely affected by the projected climatic change scenarios.

Rice Production in South Korea 11
under Current and Future Climates

JIN CHUL SHIN AND MOON HEE LEE

*Crop Experiment Station, Rural Development Administration (RDA),
Suweon, South Korea*

11.1 Introduction

South Korea is located between 33° and 38°N latitude and between 124° to
132° E longitude. The population is about 42 million (1989) and is increasing
at the rate of 1.27% per year. South Korea is in FAO-AEZ 6 (see Chapter 7),
characterized as warm subhumid subtropical with summer rainfall. However,
the Korean Peninsula has a continental weather pattern with temperature
extremes of up to 40°C in summer and as low as –40°C in winter. Precipitation
tends to vary with topography, ranging from 800 to 1300 mm annually, more
than 60% of which is received between July and September.

In recent years, South Korea has pursued policies of rapid industrialization, with the result that the population has become increasingly urbanized.
Thirty years ago, more than 70% of the population was rural and farming was
the main occupation of the majority, but by 1990, only 16% of the population
remained in rural areas, and the main occupations had shifted to industry and
services; today, farmers number about six million, or less than 15% of the population. Nevertheless, most of the urban people have some rural ties, either having come from the rural area themselves or having relatives in the countryside.
Agriculture, therefore, remains important politically and socially, even though
its contribution to the gross domestic product is less than 10%.

Rice is the staple food of South Korea, having been grown in the region for
more than 3000 years, and is still more important than any other food crop or
agricultural commodity. South Korea is self-sufficient in rice, even experiencing surpluses, although it produces only 35% of its domestic demand for other
cereals. Although the national average yield has been stable during the last ten

years at around 6 to 6.5 t ha^{-1}, both per capita consumption and the area sown of rice is gradually decreasing, the latter mainly as a result of industrialization and expanding urban areas.

Except for the south-west part of the country, most of South Korea's rice-fields lie in valleys between mountains with steep slopes. Previously, double-cropping of rice with barley or wheat was common practice, but a move towards importation of these other crops has resulted in a shift to single crops of rice. Use of modern varieties is widespread, particularly in the lowland areas. National rice yields vary strongly from year to year as a result of climatic variation. Drought occurs roughly every three years, especially during seedling growth and transplanting, but severe damage is minimized in most areas by the availability of irrigation water. More than 70% of the ricefields have stable irrigation systems; although the rest are rainfed, most have groundwater pumping facilities. Excessive rainfall from July to September can cause flooding damage, particularly in the steep areas, where soil and stones from destroyed stream banks frequently cover the ricefields because of the sharp gradient of the streams. Continuous rainfall in early September can also cause rice grains to germinate in the panicles.

At present, the government limits farm size to three hectares per household to protect smallholders, but is considering increasing this limit and rearranging land into larger blocks for eventual mechanization. However, agriculture is deeply rooted in the political and social life of the country and changes in the structure of agriculture are sensitive issues. Moreover, in the mountain areas full mechanization is difficult because of the steepness of the slopes; the size of the valleys also places an upper limit on the size of the blocks.

Production constraints include salt water intrusion in the coastal rice-fields, and nutrient deficiencies which are experienced nationwide, especially of minor elements such as iron, magnesium, zinc, and boron. Soil amendments such as foreign clay soil, zeolite, limestone, and calcium silicate are used to help alleviate these problems.

11.2 Agroecological Zones and Current Rice Production in South Korea

11.2.1 Agroecological zones

South Korea is divided into two main agricultural regions by the Taebaek mountains, one facing the East Sea which is a part of the Pacific Ocean, and the other facing the Yellow Sea. Agroecological zones (AEZs) within the country are determined mainly by elevation and latitude, although soil characteristics such as salinity, and wind from the Pacific Ocean in east coastal areas, are also important. Using these criteria, five AEZs can be distinguished (Fig. 11.1). AEZ I

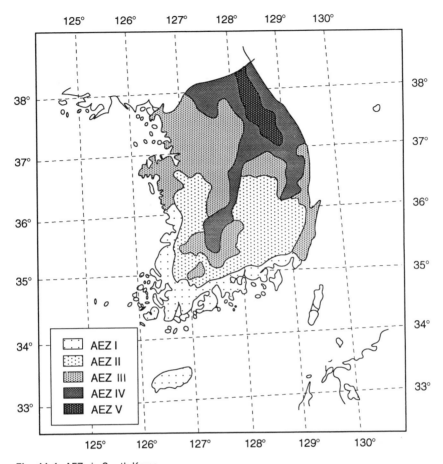

Fig. 11.1. AEZs in South Korea.

is located in the south and west coastal plain area and consists mainly of reclaimed paddy fields. AEZ II is the southern plain area where the elevation is less than 250 m, and experiences similar temperatures to AEZ I. AEZ II is further divided into two subzones differing in elevation: II–1 is lower than 150 m, II–2 is the area between 150 to 250 m of elevation. AEZ III, which is the major AEZ, is located in the northern part of the country, and also consists of a plain area below an elevation of 250 m. However, the annual mean temperature is about 1 to 2°C lower than in AEZ II. This zone is further divided into four subzones: III–1 is the plain area in which the elevation is lower than 150 m, III–2 is the area with an elevation between 150 and 250 m, III–3 is the west coastal area in which the paddy soil is saline, while III–4 is the south east coastal zone where strong cool winds from the Pacific Ocean sometimes cause problems. AEZ IV is the mid-mountainous area between 250 to 450 m elevation where

Table 11.1. Cultivated rice areas and other characteristics by province and AEZ in South Korea in 1980 (RDA, 1981).

Province	AEZ	Paddy area (ha)	Non-frost days (day)	Annual mean temp. (°C)	PPT (mm)	Weather station
Kyeonggi-do	III-1	60,063	208	11.0	1215	Suweon
	III-2	46,390	187	11.4	1135	–
	III-3	47,621	181	11.3	1282	–
	IV-1	23,521	158	10.9	1149	–
Kangweon-do	III-1	20,586	190	10–11	c.1150	Chuncheon
	III-2	9,640	153	9–10	c.1300	–
	IV-1	18,349	178	8–10	c.1050	Cheolweon
	IV-2	15,364	254	12.0	c.1250	Kangneng
	V	2,354	122	6–8	c.1250	–
Chungbuk-do	III-1	38,248	171	11.6	1216	Cheongju
	III-2	13,795	–	–	–	–
	IV-1	9,610	163	10.1	1274	–
Chungnam-do	III-3	8,594	–	–	–	–
	III-1	154,928	–	12.1	1360	Taejeon
	III-2	10,795	–	11.3	–	–
	IV-1	2,658	–	–	–	–
Cheonbuk-do	I	4,661	–	–	–	–
	II-1	128,099	–	12.9	1290	–
	II-2	20,874	–	12.9	1296	Cheonju
	III-2	14,504	–	12.2	1271	–
	IV-1	2,021	–	12.1	1173	–

Province	AEZ						Site
Cheonnam-do	I	11,222			—	c.1300	—
	II–1	162,213			13.2	c.1300	Kwangju
	II–2	31,865			—	1300–1500	—
	III–2	2,408			—	1300–1500	—
	IV–1	137			—	over 1500	—
Kyeongbuk-do	II–1	98,553			—	905	—
	II–2	59,180			13.3	1029	Chilgok
	III–4	18,534			12.7	1016	—
	IV–1	32,705			—	1054	—
	V	2,974			—	—	—
Kyeongnam-do	—	1,434			—	—	—
	III–4	30,494		208	13.5	1318	—
	II–1	95,822		202	14.0	1403	Chinju
	II–2	31,751		195	13.1	1281	Milyang
	III–2	12,287		184	11.5	1224	—
	IV–1	5,044		—	10.8	1270	—

Table 11.2. Crop calendar in each AEZ in Korea.

AEZ	Sowing			Transplanting			Heading			Harvesting
	Early limit	Optimum	Late limit	Early limit	Optimum	Late limit	Early limit	Optimum	Late limit	Late limit
I	1 Apr	20 Apr	30 Apr	27 Apr	30 May	15 Jun	2 Jul	7/25–8/30	5 Sept	20 Oct
II	1 Apr	20 Apr	30 Apr	27 Apr	30 May	15 Jun	2 Jul	7/25–8/30	5 Sept	20 Oct
III	1 Apr	15 Apr	20 Apr	10 Apr	25 May	5 Jun	12 Jul	7/17–8/30	4 Sept	14 Oct
IV	1 Apr	5 Apr	15 Apr	15 May	20 May	25 May	24 Jul	7/29–8/15	20 Aug	28 Sept
V	1 Apr	5 Apr	10 Apr	18 May	20 May	25 May	21 Jul	7/26–8/3	5 Aug	20 Sept

Table 11.3. Time course of rice yield in Korea represented by provinces from 1980 to 1990 (rough rice, t ha^{-1}). Source: Korean Ministry of Agriculture, Forestry and Fisheries (1992).

Province	Ref. weather station	1981	1982	1983	1984	1985	1986	1987	1988	1989	1990
Kyeonggi	Suweon	5.42	5.81	5.68	6.07	5.77	6.11	5.62	6.08	6.37	5.41
Kangweon	Chuncheon	4.96	5.18	5.18	5.84	5.87	5.78	5.80	5.49	6.34	5.37
Chungbuk	Cheongju	5.40	5.96	6.17	6.50	6.58	6.42	6.50	6.59	6.40	5.92
Chungnam	Taejeon	5.96	6.41	6.51	6.95	6.88	6.73	6.24	7.00	6.80	6.52
Cheonbuk	Cheonju	6.36	6.60	6.49	6.90	6.85	7.06	6.68	7.28	7.07	6.88
Cheonnam	Kwangju	5.49	5.99	6.06	6.74	6.67	6.35	5.87	7.03	6.25	6.52
Kyeongbuk	Chilgok	5.51	5.73	5.91	6.35	6.24	6.30	6.15	6.21	6.39	6.23
Kyeongnam	Chinju	5.48	5.28	5.60	6.08	5.68	5.84	5.47	6.02	5.87	6.05
Average		5.64	5.94	6.02	6.49	6.36	6.38	6.02	6.56	6.45	6.21

Table 11.4. Changes in rice area planted in Korea. Source: Korean Ministry of Agriculture, Forestry and Fisheries (1992).

Province	1980	1981	1982	1983	1984	1985	1986	1987	1988	1989	1990
Kyeonggi	181,870	179,251	178,883	179,462	178,541	180,148	179,104	188,811	185,064	186,213	180,741
Kangweon	55,987	55,288	55,661	55,778	55,849	55,800	55,406	61,990	61,049	60,841	60,545
Chungbuk	76,924	77,780	76,649	78,895	79,447	79,178	78,394	77,250	76,504	76,125	75,683
Chungnam	173,187	173,598	172,728	178,052	180,162	181,581	184,414	184,646	188,091	187,416	187,059
Cheonbuk	166,899	166,555	166,530	169,166	170,574	174,471	175,186	172,844	175,874	176,743	176,385
Cheonnam	208,404	208,319	207,002	210,335	211,516	213,866	216,373	214,580	215,780	215,508	215,130
Kyeongbuk	199,050	195,320	171,275	192,966	194,252	193,493	190,769	198,031	194,096	193,231	190,844
Kyeongnam	157,518	156,147	147,016	154,961	154,134	154,387	153,033	160,989	159,700	158,154	155,116
Total	1,219,839	1,212,258	1,175,944	1,219,615	1,224,475	1,232,924	1,232,679	1,259,141	1,256,158	1,254,231	1,241,503

chilling injury of rice is frequently observed, and is further divided into two subzones: IV–1 is the mid-mountainous which is located along the Taebaek mountains, while IV–2 is located in the north-east coastal area. AEZ V is the high mountainous area where rice does not grow well. Table 11.1 shows the area distribution of the AEZ zones in each province and their climatic characteristics. About 90% of the area planted to rice in South Korea is in AEZ II and AEZ III; only 1.7% of the area is in AEZ I and AEZ V together.

11.2.2 Rice cropping calendar

Table 11.2 shows the crop calendar by AEZ. Transplanting with a rice transplanter is common in Korea. Rice is sown in seedling boxes (60 × 30 × 2.5 cm height) and germinated under incubator conditions and raised in a nursery covered by polyethylene film to avoid low temperature damage. Optimum sowing dates are different in each AEZ depending on the air temperature. The date of sowing is determined by the earliest possible transplanting date when temperatures have to be higher than 13.5°C. These temperatures are needed for root development of seedlings. Seedlings are normally transplanted at 35 days of age. The latest possible transplanting date depends on the temperature during ripening, a process that stops below minimum temperatures of 10°C. To complete maturation, heat units of about 640°Cd (with a base temperature of 8°C) after heading are required.

11.2.3 Historical rice production

Table 11.3 shows the yield variability within each province and the national average yield between 1981 and 1990. The lowest yield was observed in 1981 and highest yield in 1988. Yields are lowest in the provinces Kangweon-do and Kyeongnam-do. In the case of Kangweon-do, the main rice-growing areas are located in AEZ III–3 or IV, where low temperatures determine yield potential. In Kyeongnam-do, the main rice-growing area is located in AEZ II where a double-cropping system is applied. Because rice is planted late, cool temperature problems at the end of the growing season may also occur. Table 11.4 shows that the area planted with rice has gradually decreased during the past ten years.

11.3 Model Parameterization and Testing

The ORYZA1 rice potential production model described in Chapter 3 was used in the present study to evaluate the effect of climate changes on rice production in South Korea.

11.3.1 Parameterization of ORYZA1 for Korean varieties and climatic conditions

The ORYZA1 model was initially developed using experimental data from the *indica* variety IR72 grown at IRRI in the Philippines. However, the model is able to simulate the growth and yield of other varieties in other environments by appropriate calibration of its parameters. For this, data was used for the *japonica* variety Hwaseongbyeo grown at 165 kg N ha^{-1} in an experiment at the Crop Experiment Station in 1992, Suweon, South Korea (CES, 1992a). Hwaseong-byeo is moderately sensitive to photoperiod, and differs from IR72 mainly in the proportions of dry matter allocated to the different plant organs. It was also found that the initial relative growth rate of the leaves (RGRL) was higher in Hwaseongbyeo than IR72. A value of 0.015 was calculated from the experimental data, which is somewhat higher than the 0.007 used for IR72. The model was also adapted to take into account the fact that during the nursery period the seedlings are generally covered by polyethylene film to avoid low temperature damage, although if the temperature under the film rises higher than 25°C, the polyethylene cover is usually opened. In the model, this practice was accounted for by assuming that the daily maximum and minimum temperatures inside the polyethylene film tunnel were respectively 3.5°C and 2.5°C higher than the corresponding ambient temperatures, but if this corrected temperature exceeded 25°C, then normal ambient air temperature was used. Other modifications made to the model included the death rate of the leaves as a function of development stage (leaf death rate in Hwaseongbyeo is remarkably slower than IR72). Other parameters remained the same as for IR72 at IRRI. Values of the genotypic parameters used in the model are shown in Table 11.5.

With these changes, the model simulated the leaf area index and the time course of dry matter growth of the individual organs accurately for the 1992 data set from which the model was calibrated and a data set from 1991 from an experiment conducted at the same site (Fig. 11.2).

11.3.2 Model performance

The model was tested firstly with data from regional yield tests conducted in 1990 by RDA (CES, 1992b) using the two varieties Dongjinbyeo and Chucheongbyeo. Table 11.6 shows the data used for validation. Management characteristics such as transplanting date, number of plants per hill, planting density were used as input to the model. As the two varieties Dongjinbyeo and Chucheongbyeo are similar to the variety Hwaseongbyeo used for calibration, all parameters except for those describing phenology were assumed to be the same; recorded heading dates were used to calculate the phenological parameters. The simulation results are shown in Fig. 11.3. Dongjinbyeo, which

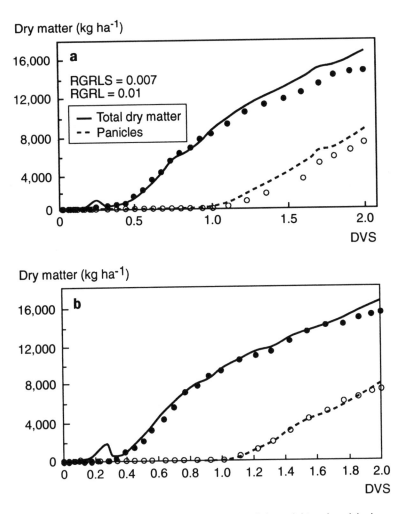

Fig. 11.2. Simulated (lines) and observed (symbols) total dry weight and panicle dry weight in relation to development stage (DVS) for 1992 (a) and 1991 (b) experiments.

was grown in AEZ II, showed hardly any variation in observed yield but a significant variation in yield was simulated based on the climatic data. Yields of the variety Chucheongbyeo, which is grown in AEZ III, varied among sites and observed yields were closely related to the simulated yield ($r = 0.78$). However, the relationship differed from the 1:1 line, indicating that the model was simulating a greater variation in yields between the different environments than was actually observed. This may be due to the overriding effect of other environmental (e.g. soil type) or management factors that limit yield besides temperature and solar radiation.

Table 11.5. Genotype coefficients and function tables (expressed as a function of developmental stage, DVS) for the variety Hwaseongbyeo used in the simulations. FSHTB, FLVTB, and FSTTB are the fractions of new assimilate partitioned to the shoot, leaves and stem respectively; SLATB is the specific leaf area (cm g^{-1}), DRLVT is the relative death rate of the leaves (g g^{-1}), and NFLVTB is the leaf nitrogen concentration (g N m^{-2} leaf).

Genotype coefficient	Name	Units	Value
Initial relative leaf area growth rate	RGRL	$(°Cd)^{-1}$	0.015
Stem reserves fraction	FSTR	–	0.4
Basic vegetative period duration	JUDD	Dd	20.0
Photoperiod sensitive phase duration	PIDD	Dd	15.0
Minimum optimum photoperiod	MOPP	h	11.5
Photoperiod sensitivity	PPSE	Dd h^{-1}	0.2
Duration of panicle formation phase	REDD	Dd	25.0
Duration of grain-filling phase	GFDD	Dd	33.2
Maximum grain weight	WGRMX	mg $grain^{-1}$	25.0
Spikelet growth factor	SPGF	sp g^{-1}	60.0

FSHTB = 0.0, 0.50, 0.305, 0.759, 0.333, 0.684, 0.389, 0.821, 0.462, 0.833, 0.547, 0.970,
        0.640, 0.941, 0.737, 0.938, 0.827, 0.979, 1.018, 1.000, 2.050, 1.000

FLVTB = 0.0, 0.45, 0.305, 0.445, 0.333, 0.315, 0.389, 0.357, 0.462, 0.361, 0.547, 0.393,
        0.640, 0.391, 0.737, 0.390, 0.827, 0.193, 0.950, 0.000, 2.050, 0.000

FSTTB = 0.0, 0.55, 0.305, 0.555, 0.333, 0.685, 0.389, 0.643, 0.462, 0.639, 0.547, 0.607,
        0.640, 0.609, 0.737, 0.610, 0.827, 0.807, 0.950, 0.950, 1.000, 0.900, 1.200,
        0.000, 2.050, 0.000

SLATB = 0.0, 368, 0.305, 368, 0.333, 305, 0.389, 324, 0.462, 305, 0.547, 270, 0.640,
        280, 0.737, 273, 0.841, 277, 0.953, 276, 1.084, 272, 1.231, 256, 1.380, 251,
        1.534, 243, 1.689, 249, 1.825, 252, 1.937, 232, 1.992, 223, 2.05, 223

DRLVT = 0.0, 0.0, 0.6, 0.0, 0.95, 0.0, 1.0, 0.008, 1.6, 0.008, 2.1, 0.015

NFLVTB = 0.0, 0.487, 0.305, 0.815, 0.333, 1.201, 0.389, 1.270, 0.547, 1.408, 0.737,
        1.318, 1.231, 0.986, 1.534, 0.820, 1.825, 0.628, 1.992, 0.547, 2.05, 0.51

The capacity of the model to simulate variation in yield across years was evaluated with data from an experiment conducted from 1980 to 1989 by RDA with the two varieties M15 and Kwanakbyeo. M15 was used in Yongnam Crop Experiment Station for ten years and Kwanakbyeo was used in several sites in AEZ IV. Yields were simulated using the leaf nitrogen concentration that was measured in an experiment at RDA. The variation in observed yield was higher for M15 than for Kwanakbyeo (Fig. 11.4). The correlation coefficient for the relationship between simulated yield and actual yield of M15 was significant ($r = 0.829$). The results demonstrate that the model is capable of simulating relative yield differences resulting from climatic variability between years.

The model was subsequently used to predict long term yield trends in six provinces in South Korea and compared with recorded farmers' yields. The

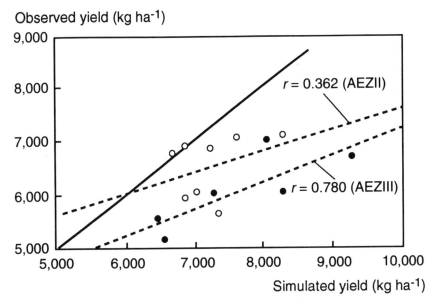

Fig. 11.3. Simulated and actual yield of Dongjinbyeo (open symbols) in AEZ II and Chucheongbyeo (closed circles) in AEZ III.

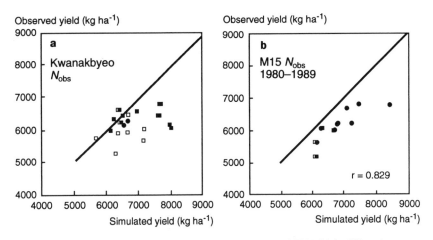

Fig. 11.4. Actual and simulated yields of Kwanakbyeo (a) and M15 (b) in different years.

crop calendar used for the simulation is given in Table 11.2. Weather data from three stations in each province were used for the simulations. Simulated and recorded farmers' yields for the different provinces in the period from 1981 to 1990 are shown in Fig. 11.5. The simulated potential yields were only slightly

Table 11.6. Data used for the validation of model.

Year	AEZ	Site	Variety	Sowing	Transpl.	Heading	Density	Plants per hill	Yield
1990	II–1	Iri	Dongjinbyeo	20 Apr	30 May	14 Aug	22.2	3	6,750
1990	II–1	Taejeon		20 Apr	30 May	16 Aug	22.2	3	7,010
1990	II–1	Milyang		20 Apr	30 May	16 Aug	22.2	3	6,990
1990	II–1	Milyang		1 Jun	1 Jul	26 Aug	27.8	5	6,400
1990	II–1	Chilgok		20 Apr	30 May	13 Aug	22.2	3	6,800
1990	II–1	Chilgok		15 May	15 Jun	21 Aug	27.8	5	6,850
1990	II–1	Chinju		20 Apr	30 May	14 Aug	22.2	3	5,640
1990	II–1	Chinju		15 May	15 Jun	20 Aug	27.8	5	5,910
1990	III–1	Suweon	Chucheongbyeo	15 Apr	25 May	23 Aug	22.2	3	5,980
1990	III–1	Suweon		25 May	25 Jun	1 Sep	27.8	5	5,560
1990	III–1	Chuncheon		15 Apr	25 May	26 Aug	22.2	3	5,160
1990	III–1	Cheongju		15 Apr	25 May	20 Aug	22.2	3	6,980
1990	III–4	Yuongduk		15 Apr	25 May	21 Aug	22.2	3	6,640
1990	III–4	Yuongduk		15 May	15 Jun	28 Aug	27.8	5	5,970

higher than farmers' yields, indicating that farmers manage their crops inten-
sively. Year to year variability was small for both the simulated and recorded
yields.

11.4   Effect of Climate Change on Rice Production in South Korea

11.4.1   Input data

The ORYZA1 model, calibrated for the variety Hwaseongbyeo as described
above, was used to simulate potential rice yields under various scenarios of
changed climate, including 'fixed-scenario' changes of CO_2 only (1.5 and 2

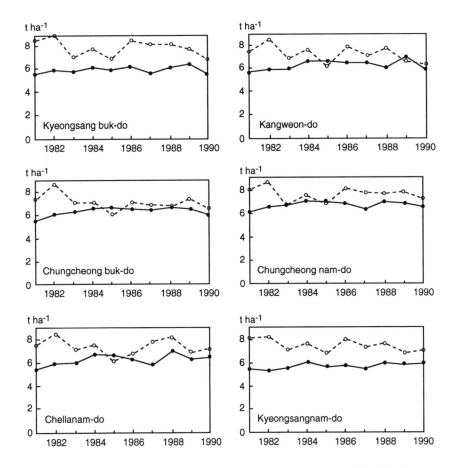

Fig. 11.5. Actual farmers' yields (solid lines) and simulated yields (dashed lines) in six
provinces in South Korea.

times the current level), temperature only (+1°C, +2°C, and +4°C above current temperatures), and under scenarios predicted by the GFDL, GISS, and UKMO General Circulation Models (GCMs) described in Chapter 5. In all, 15 different scenarios were simulated, details of which are given in Table 7.5. Daily weather data for eleven weather stations were used, details of which are given in Chapter 2. At least 12 years' weather data (and for some up to 18 years') were available for each station (Table 2.1). Table 11.7 shows the AEZ and

Table 11.7. Details of the weather stations in South Korea used in the climate change analysis. See also Table 2.1 for further details of each station.

				Julian date of	
ID	Station	AEZ	Province	Sowing	Transplanting
1	Milyang	II	Kyeongnam-do	110	150
2	Suweon	III	Kyeonggi-do	105	145
3	Cheongju	III	Chungbuk-do	105	145
4	Chilgok	II	Kyeongbuk-do	110	150
5	Chinju	II	Kyeongnam-do	110	150
6	Chuncheon	III	Kangweon-do	105	145
7	Taejeon	II	Chungnam-do	110	150
8	Kangneng	IV	Kangweon-do	95	135
9	Kwangju	II	Cheonnam-do	110	150
10	Cheonju	II	Cheonbuk-do	110	150
11	Cheolweon	IV	Kangweon-do	95	135

the province into which each station falls, and the sowing and transplanting dates used in the simulations. Six of the stations were in AEZ II, three in AEZ III, and two in AEZ IV. These three AEZs together represent about 98% of the cultivated area of rice in South Korea. The recorded daily values of minimum and maximum temperatures were adjusted by the changes used for each scenario, and in the case of the three GCM scenarios, solar radiation was also adjusted by the predicted factor. The current CO_2 concentration level was assumed to be 340 ppm, so that the $1.5 \times CO_2$ scenario represented 510 ppm and the $2 \times CO_2$ scenario represented 680 ppm.

The leaf nitrogen concentration measured in the variety Hwaseongbyeo in the experiment described above was used for all the simulations. It is recognized that under higher levels of CO_2, the leaf nitrogen concentration may be 'diluted', thereby reducing photosynthetic rate per unit leaf area; on the other hand, however, the increased growth under higher CO_2 may result in a larger root system, so that N uptake is also increased. In the absence of reliable quantitative data on this effect, therefore, we have assumed that leaf N concentration does not change.

11.4.2 Effect of fixed increments of temperature and CO_2 on potential yields

Results of the simulations are shown for each AEZ in Table 11.8. Values are the average percentage changes in potential yields of all the years and all the stations in each zone. Increasing the CO_2 concentration increased the yield in all

Table 11.8. Predicted changes in potential rice yields in South Korean AEZs II, III, and IV with changes in CO_2 level only, temperature only, and under scenarios predicted by the GFDL, GISS, and UKMO GCMs.

AEZ	Effect of CO_2		Effect of temperature			Effect of GCM		
	$1.5 \times CO_2$	$2 \times CO_2$	+1°C	+2°C	+4°C	GFDL	GISS	UKMO
II	20.2	31.4	−11.0	−19.7	−35.0	−12.8	−3.1	−32.2
III	18.6	28.9	−7.5	−12.5	−23.3	0.0	9.5	−11.3
IV	16.2	24.6	−5.9	−10.8	−18.2	6.1	11.8	−0.6

zones by an average of 18.3% and 28.3% for the $1.5 \times CO_2$ and $2 \times CO_2$ scenarios, respectively. The effect, however, was less marked in AEZ III and AEZ IV, which are at higher altitudes than AEZ II. Increasing temperatures decreased the yield in all zones, but again the effect was less in the higher altitude zones. Average decreases in yields in AEZs II, III, and IV were −9.9%, −6.5%, and −5.3% per 1°C increase, respectively.

11.4.3 Effect of predicted GCM scenarios on potential yields

Of the three GCM scenarios, the UKMO predicted the greatest decline in yields, ranging from −0.6% in AEZ IV down to −32.2% in AEZ II. The least detrimental scenario was the GISS, under which yield increases were predicted for two of the zones, and only a −3.1% decrease for AEZ II. The GFDL scenario was intermediate between these two extremes. More interestingly, however, is the pattern of changes in each AEZ with respect to each other. In all scenarios, the greatest declines in yield are in the lower altitude AEZ II, with the least detrimental effects in AEZ IV. These results suggest that under a changed climate, there could be a shift in rice production in South Korea to higher altitudes.

It is interesting to look at the factors causing this relative increase in yields at higher altitudes in more detail. For this, the three stations Milyang, Cheongju, and Cheolweon were chosen as representative of AEZs II, III, and IV, respectively. The stations ranged in altitude from 12 m to 192 m above sea level (Table 11.9). Simulations were made for a single year for each station for the

Table 11.9. Simulated potential yields under the GFDL scenario for representative stations in the three South Korean AEZs used in the study. The GFDL scenario was taken as representative of the average of all three GCMs (see Table 11.8). A doubled-CO_2 level of 680 ppm was used with the scenario.

Station	AEZ	Altitude (m)	Spikelets at anthesis (m^{-2})	Filled grain fraction	Filled grains (m^{-2})	Yield (kg ha^{-1})	Crop duration (days)	Temp. at flowering (°C)
Milyang	II	12	54,301	0.251	13,634	3,963	136	37.9
Cheongju	III	59	49,144	0.723	35,553	10,335	139	35.5
Cheolweon	IV	192	61,334	0.954	58,513	17,009	143	31.3

doubled-CO_2 GFDL scenario only, the results of which are shown in Table 11.9. The simulated potential yields ranged from 4 to 17 t ha^{-1}, most of which variation was explained by large differences in the filled grain fraction, ranging from 25% to 95%. Examination of the mean temperatures around the time of flowering showed that at the lower altitudes, mean maximum temperatures of almost 38°C were responsible for this high spikelet sterility. At higher altitudes (AEZ IV), the temperatures at the time of flowering were considerably lower. As in other chapters in this book, therefore, spikelet sterility seems to be a major factor limiting yields under the higher temperatures of changed climates. Selection for more temperature tolerant varieties may be possible to offset this detrimental effect, particularly as there appears to be considerable genotypic variation in the sensitivity of spikelet fertility to high temperatures (see Chapter 7).

It is useful to gain some idea of the magnitude of the effect of these predicted changes on overall rice production in South Korea. As current annual production figures were not available on an AEZ basis, those for each province were used instead to stratify overall national production (Table 11.10). Current production for each province was adjusted by the predicted change for the province (mean of all stations and years) and summed to give the overall effect on national production. In this way, changes of –9%, +1% and –25% were predicted for the GFDL, GISS, and UKMO scenarios respectively. Thus, the predicted effect of climate change depends very much on which GCM scenario is used.

11.5 Conclusions

The effect of climate change on national annual rice production depended on the scenario used, ranging from –25% to +1%. The average of these would suggest a decline in production of about 11%. The main reason for this was high summer temperatures causing significant spikelet sterility. However,

Table 11.10.   Predicted effect of climate change under the three doubled-CO_2 GCM scenarios on overall national production of rice in South Korea, stratified according to provinces. Production figures (tonnes, t) in 1990 are used as the current baseline, and are adjusted by the predicted fractional changes for each province. Predicted changes are the means of all years for all weather stations in each province (see Table 11.7).

Province	Current (t)	GFDL % incr.	(t)	GISS % incr.	(t)	UKMO % incr.	(t)
Kyeonggi-do	995,883	2.8	1,023,604	12.0	1,115,876	1.2	1,007,961
Kangweon-do	325,127	2.9	334,611	9.6	356,179	−6.5	303,984
Chungbuk-do	448,043	0.8	451,658	11.6	500,119	−16.5	373,944
Chungnam-do	1,219,625	−9.2	1,107,970	0.0	1,219,061	−27.5	884,775
Cheonbuk-do	1,206,649	−11.8	1,064,533	2.1	1,231,790	−30.6	837,799
Cheonnam-do	1,402,648	−9.0	1,277,074	−0.4	1,396,478	−24.9	1,053,046
Kyeongbuk-do	1,188,958	−23.9	904,692	−14.9	1,012,278	−44.7	657,132
Kyeongnam-do	938,452	−8.8	855,644	0.1	939,789	−29.8	658,594
Total	7,725,385		7,019,786		7,771,570		5,777,235
Overall change (%)			−9.1		0.6		−25.2

although the magnitude of the change varied, all GCMs predicted a greater relative effect on rice production at lower altitude AEZs. This has implications for planners in the next century; if rice production is to shift gradually to higher altitude areas, further development of infrastructure may be necessary to transport rice from the areas of production to those of consumption. There is also work for rice breeders; selection of varieties with greater temperature tolerance of spikelets may help to maintain rice production at current levels.

Rice Production in China under Current and Future Climates 12

Zhu Defeng and Min Shaokai

China National Rice Research Institute, Hangzhou 310006, China

12.1 Introduction

China, the world's most populous country, is situated between latitudes of 18° and 54° N and longitudes of 73° and 135° E. In 1990, the population was nearly 1.14 billion, increasing at a rate of 1.56%, with the agricultural population accounting for 66% of this total. About 75% of the population is concentrated in the northern, north-eastern, eastern and south central areas, which make up only 44% of the nation's land area. The remaining 25% of the population is dispersed in the south-western and north-western parts, which account for 56% of the total land area. This gradually decreasing population density to the west reflects the heterogeneity both in regional natural features as well as in socioeconomic development.

Because of its size, China spans four FAO-AEZs (see Chapter 7), all of which are subtropical and include some temperate areas. These are: FAO-AEZ 5, the warm arid and semi-arid subtropics with summer rainfall; FAO-AEZ 6, the warm subhumid subtropics with summer rainfall; FAO-AEZ 7, the warm/cool humid subtropics with summer rainfall; and FAO-AEZ 8, the cool subtropics with summer rainfall. Rice is produced primarily in FAO-AEZs 6 and 7.

Agriculture contributes about 27% to China's gross domestic product which totaled $364.9 billion in 1990. Rice is the staple food of China. Although the rice cropping area represents only 29.1% of the total national crop-growing area, rice production contributes 43.7% of total national grain production, representing 22.8% and 36.9% of the total world cropping area and production respectively (Xiong *et al.*, 1992). Of China's total production, which has ranged from about 171 million to 191 million tonnes annually for the past decade and is increasing at an average of 2.8% annually, less than 1% enters world trade.

Because the domestic rice requirement will continue to rise over the long term, China concentrates on increased rice production to meet the growing needs of its population. However, a shrinking land base for agriculture, both on a per capita and absolute basis, is the primary production constraint. On average, China's population is increasing by 17 million year^{-1} while cultivated land is shrinking by some 300,000 ha year^{-1}. In the last 40 years, cultivated land per capita has shrunk from 0.18 ha to 0.085 ha, a decrease of 53%. In addition, water loss, soil erosion, expanding desertification and salinization, and decreasing soil organic matter contents pose serious obstacles for continued agricultural development. The increasing demand for rice, therefore, has to be met mainly by raising yields; some 26% of total rice area, which currently has medium-to-low productivity, offers some scope for this.

Rice cultivation in China is distributed across a vast area which spans temperate, subtropical, and tropical belts (Xiong *et al.*, 1992; IRRI, 1993), with the greatest production from the subtropical belt. The distribution of rice cropping is characterized by a gradual decrease from the south to the north, with the greatest concentration in the south-east. Small, scattered areas of production occur in the north-west (CNRRI, 1988; Xiong *et al.*, 1992; IRRI, 1993, pp. 46–48). The northern-most region of rice production is at Mohe in Heilongjiang Province, at a latitude of 53.3°N.

About 93% of the total rice cultivation in China is paddy-irrigated. Although conditions for rice growth are generally favorable throughout China, in some areas, extreme events such as floods, droughts, typhoons, and adverse low and high temperatures can severely limit the efficiency of rice production.

Precipitation during the cropping season varies considerably between different rice-growing regions. There is a general trend for the distribution of precipitation to decrease from south-east to north-west. The lower precipitation in the north, particularly with the occurrence of spring drought, is a major constriction on the expansion of the rice cultivation area, and significant reductions in yield often occur due to a delay in transplanting as the farmers await the arrival of the rains. In the south, drought occasionally results from large variations in the distribution and amount of annual rainfall. In Sichuan, Hubei, and Jiangxi Provinces, where rice is double-cropped, drought sometimes affects the grain-filling stage of the first crop ('early' rice), or the recovering stage after transplanting of the second crop ('late' rice). Autumn drought can reduce yields in the eastern and southern hill regions by affecting the grain-filling period of 'single' crops, or the heading and flowering stages of late rice in double-cropped systems (Min & Fei, 1984). Excessive rainfall can affect tiller formation and panicle development of both early rice and single rice through the reduction in solar radiation, and may even cause damage from flooding in south-east regions. Typhoons can also seriously affect rice production by causing lodging, as well as floods in the south-east areas.

Adverse temperatures also have a major influence on rice production in

China. High temperatures often occur during the heading and flowering stages of single rice and early rice, decreasing yields by increasing panicle sterility. Low temperatures also impose an important constraint on rice yields throughout the country by causing cold damage to young seedlings early in the season. This is particularly so in southern China during January and February, in central China during March and April, and in northern China during April and May. Low temperatures can also affect meiosis during panicle development in many regions. Large areas of rice cultivation also suffer low temperatures during flowering, particularly in the Yangtze river valley (September), in south China (mid-September to early October), and in northern and north-east China (mid-August). Although low temperatures can occur throughout the growing season, it is only when they coincide with flowering that the greatest damage is done (Min & Fei, 1984).

The climate, therefore, has a major influence on rice production in China. Any changes in the current climate, such as those predicted to occur due to an increased level of greenhouse gases in the atmosphere, are likely to have a profound impact on national rice production, either positively or negatively. It is, therefore, vitally important to assess the impact on regional rice yield so as to provide a basis for policy making, cultivar improvements, and alteration of cropping seasons and cultivation technology.

12.2 Agroecological Zones and Current Rice Production in China

12.2.1 Agroecological zones

The wide distribution of areas in which rice is grown in China also means that there is a wide range of environments to be considered. An analysis of ambient temperatures in different rice systems indicates that single-cropped rice is grown where 2000–4500°Cd of annual accumulated temperature (above a base temperature of 10°C) can be achieved, double-cropped rice where 4500–7000°Cd can be achieved, while triple-cropped rice is possible where the accumulated thermal time exceeds 7000°Cd. A recent study has suggested that 5300°Cd should be taken as the limit for double-cropped rice (CNRRI, 1988). Based on differences in ecological and social conditions between different regions, rice cultivation in China has been classified into six zones and sixteen subzones (CNRRI, 1988). These are summarized in Fig. 12.1 and Table 12.1. The climate and cropping systems in each zone, and the areas of rice cultivation and the production, are summarized in Tables 12.2 and 12.3, respectively.

The largest areas of rice production are in Zones I, II and III, with Zone III representing 93.6% of the total rice cropping area in China. The smallest area

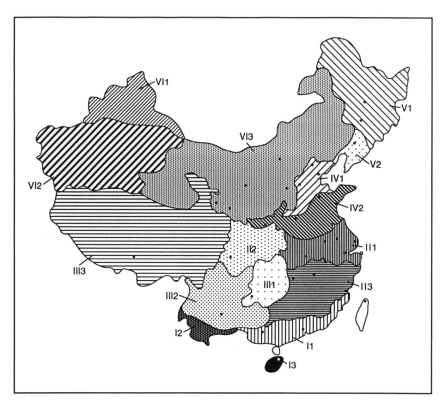

Fig. 12.1. The main AEZs of China (refer to Table 12.1 for description).

of rice cultivation is in Zone VI with only 0.5% of the total. Yields of single-cropped rice, on the other hand, are highest in Zones V and IV.

12.2.2 Rice cropping calendar

Going from south to north, there is a decrease in the total amount of solar radiation received, average temperatures, and amount of rainfall during the growing season. Due to cold winters, the length of the period suitable for growing rice is also shorter in northern regions (Xiong *et al.*, 1992). The duration of rice crops in various cropping seasons of each zone is shown in Table 12.4. In general, sowing and maturity dates become later moving from south to north. Single-cropped rice has a longer growth duration.

Typical dates of sowing and transplanting in each agroecological zone (AEZ) are shown in Table 12.6. In general, single rice is transplanted at 35–40 days, early rice is planted at 25 days, and late rice at 30 days after sowing.

Table 12.1.   Description of AEZs in China.

Zone	Sub-zone	Description
I		South China (double rice cropping)
	1	The subzone of double rice cropping in the Fujian-Guangdong-Guangxi-Taiwan plain and hilly areas
	2	The subzone of single rice cropping in the South Yunan river valley and basin
	3	The subzone of multiple cropping for double rice crops in the Hainan-Leizhou peninsula tableland and plain
II		Central China (double and single rice cropping)
	1	The subzone of double and single rice cropping on the middle and lower reaches of the Yangzi River plain
	2	The subzone of double cropping for single rice crop in the Sichuan-Shanxi basin
	3	The subzone of double rice cropping in the hilly and plain areas south of the Yangzi River
III		South-western plateau (single and double rice cropping)
	1	The subzone of single and double rice cropping in the East Guizhou–West Hunan plateau mountainous areas
	2	The subzone of double cropping for single rice crop in the Yunnan–Sichuan plateau ridge valley
	3	The subzone of single rice cropping of the Qinghai-Tibet high cold river valley
IV		North China (single rice cropping)
	1	The subzone of medium and early maturing rice crop for single rice cropping on the plain of the north part of North China
	2	The subzone of medium and late maturing rice crop for single rice cropping in the Huaihe River plain and hilly areas
V		North-east China (early maturing rice cropping)
	1	The subzone of very early maturing rice cropping in the Heilongjiang-Jilin plain and river valley
	2	The subzone of early maturing rice cropping on the coastal plain of the Liaohe River
VI		Northwest China (single rice cropping region in dry areas)
	1	The subzone of early maturing rice cropping in the North Xinjiang basin
	2	The subzone of medium maturing rice cropping in the South Xijiang basin
	3	The subzone of early and medium maturing rice cropping on the Gansu-Ningxia-Shanxi-Inner Mongolia plateau

Special consideration is made for the seedbed period of early rice, the first crop in double-cropped rice, and for single rice in the northern-most sites, during which the mean ambient temperature can be as low as 10–12°C. Farmers ensure the survival of the seedlings during this period by raising them under

Table 12.2. Summary of the climate in each of the AEZs.

Zone	Annual accumulated temperature sum (°Cd)[1]	Growing season (days)	Rainfall (mm)	Annual radiation (1000 cal.)	Annual sunshine (h)
I	5800–9300	260–365	1200–2500	90–120	1500–2600
I.1	6500–8000			90–110	
I.2	5800–7000			90–100	
I.3	8000–9300			100–120	
II	4500–6500	210–260	800–2000	50–115	1200–2300
II.1	4500–5500			80–115	
II.2	4500–6000			50–90	
II.3	5300–6500			80–110	
III	2900–8000	170–210	800–1400	70–110	1200–2600
III.1	3500–5500			70–90	
III.2	3500–8000			80–110	
III.3	2900–4400				
IV	4000–5000	110–200	580–1000	110–135	2000–3000
IV.1	4000–4600		5800–6300	120–135	2400–3000
IV.2	4000–5000		600–1000	110–125	2000–2600
V	2000–3700	110–200	350–1100	100–146	2200–3100
V.1	2000–3100		400–1000	100–120	2200–3100
V.2	2900–3700		350–1100	116–146	2300–3000
VI	2000–4250	110–250	50–600	130–150	2500–3400
VI.1	3450–3700		150–220	130–145	2600–3300
VI.2	4000–4250		50	145–150	2800–3300
VI.3	2000–3600		200–600		2500–3400

[1] Annual accumulated temperature of more than 10°C.

transparent plastic covers, where the mean temperature is typically 5°C higher than outside (Fei, pers. comm.).

12.2.3   Historical rice production in China

Changes in the national rice yield and production over the period 1949–1990 are shown in Fig. 12.2. Owing to improvements in the varieties used and production technology, a steady increase in overall yields has occurred. Variations in the area sown and overall production have occurred due to both changes in governmental policy and other factors such as the profitability of rice production to the farmer.

Table 12.3. Description of the main cropping systems in each AEZ, including the rice cropping area and its production.

Zone	Main cropping system	Area (10³ ha)	Fraction of total rice area in each rice cropping system Double	Single	Average yields (kg ha⁻¹)
I		6,024.7	74	26	3,960
I.1	Double rice+WC	4,930.7	94	6	4,200
I.2	Double rice+WC	491.3	87	13	3,520
I.3	Double rice+WC	602.7	68	32	2,775
II		23,154.0	40	60	4,590
II.1	Double rice+WC	9,488.7	45–80	55–20	4,770
II.2	Single rice+WC	3,187.3	7	93	5,220
II.3	Double rice+WC	10,478.0	66–80	34–20	4,230
III		2,678.0	7	93	4,335
III.1	Single rice+WC	1,545.3	0	100	4,260
III.2	Single rice+WC or single rice	1,130.7	0	100	4,455
III.3	Single rice	2.0	0	100	3,375
IV		1,119.3	0	100	4,575
IV.1	Single rice+WC or single rice	260.7	0	100	4,800
IV.2	Single rice+WC	858.7	0	100	4,515
V		860.7	0	100	5,055
V.1	Single rice	459.3	0	100	4,155
V.2	Single rice	401.3	0	100	6,090
VI		187.3	0	100	4,365
VI.1	Single rice	28.0	0	100	3,270
VI.2	Single rice	66.0	0	100	2,965
VI.3	Single rice	93.3	0	100	5,760
Totals		34,024.0			4,470

WC; winter crop, which usually are winter wheat, barley, oil rape, green manure, and others.

Since 1970, rice cultivation in China has passed through three main stages. In the first stage from 1970 to 1976, the area of rice cultivation was extended continuously at an annual rate of 12.4%, mainly through a change from single-cropped to double-cropped rice, although yields over the same period only increased at an annual rate of 2.1%. Overall production increased annually by 14.8%. In the second stage from 1977 to 1984, yields increased by 55% but the cultivated area decreased by 7.8%, so that overall production increased by 41.7%. In the third stage from 1985 to 1990, the area sown, average yields and overall production remained relatively stable. However, it has been predicted that the area planted to rice in the future will decrease, due to the extension of infrastructure and residential areas. Maintaining the present level of production is, therefore, a challenge to scientists and policy makers.

Table 12.4. Duration (days) of rice crop by AEZ.

AEZ	Site	Season		
		Early	Late	Single
I	Guangzhou	110	110	130
II.1	Nanjing			140
II.1	Hangzhou	115	135	140
II.1	Wuhan	120	125	135
II.2	Chendu			145
II.3	Chansha	115	125	135
II.3	Fuzhou	115	132	
III	Guiyang			160
IV	Beijing			135
V	Shenyang			135

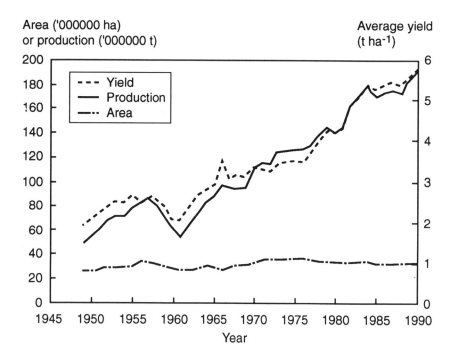

Fig. 12.2. Historical rice production statistics of China, showing the area sown, average national yields, and annual national production. (Source: IRRI, 1991.)

12.3 Model Parameterization and Testing

12.3.1 Parameterization of the crop model for Chinese varieties and conditions

The rice potential production model ORYZA1 (v1.22), described in Chapter 3, was used to evaluate the effect of change in temperature, solar radiation, and ambient CO_2 level on the potential production of rice in China.

Five varieties, Guangluai 4, Shenyou 6, Shenyou 63, Xiushui 11, and Liming, were taken to be representative of varieties grown in the main rice-growing areas and rice cropping systems of China. Guangluai 4 has been grown for about 20 years as early rice in double-cropping regions in Zones I and II, and is now being grown also at other locations. Shenyou 6 and Xiushui 11 are often grown as late rice in double-cropped systems in Zones I and II. Shenyou 63 is grown as single-cropped rice in Zones I, II, and III, while Liming represents varieties normally used in Zones IV and V. Genotype-specific parameters and functions describing development rates, dry matter partitioning, specific leaf area, and leaf nitrogen content, were derived from experimental data or literature for each of the five varieties (Table 12.5). Other, more general parameters were assumed to be the same as for the variety IR72 used in the initial development of the ORYZA1 model.

Table 12.5. Genotype coefficients and function tables (expressed as a function of developmental stage, DVS) for the five varieties used in the simulations. See Table 7.1 for a description of genotype coefficients.

Name	Units	Guangluai 4	Shenyou 6	Shenyou 63	Xiushui 11	Liming
RGRL	(°Cd)$^{-1}$	0.015	0.015	0.015	0.013	0.013
FSTR	–	0.3	0.3	0.3	0.3	0.3
JUDD	Dd	12.3	22.8	28.7	15.6	25.8
PIDD	Dd	15	15	15	12	15
MOPP	h	0	0	0	12.5	0
PPSE	Dd h^{-1}	0	0	0	0.33	0
REDD	Dd	21.6	20.8	25.8	20.0	24.0
GFDD	Dd	24.6	21.7	27.4	25.4	25.4
WGRMX	mg grain^{-1}	30.0	30.0	30.0	30.0	30.0
SPGF	sp g^{-1}	49.8	49.8	49.8	49.8	49.8

Specific leaf area (SLATB, cm^2 g^{-1})

Guangluai 4   0, 360, 0.2, 360, 0.65, 300, 1, 300, 2, 250, 2.1, 250
Shenyou 6   0, 410, 0.304, 410, 0.319, 367, 0.362, 332, 0.507, 253, 0.582, 247, 0.642, 240, 0.696, 235, 0.998, 210, 1.348, 206, 1.738, 198, 2.1, 198
Shenyou 63   0, 410, 0.304, 410, 0.319, 367, 0.362, 332, 0.507, 253, 0.582, 247, 0.642, 240, 0.696, 235, 0.998, 210, 1.348, 206, 1.738, 198, 2.1, 198

Continued

Table 12.5. Notes, *Continued*

Xiushui 11   0, 324, 0.61, 300, 0.8, 285, 1, 270, 1.55, 250, 2, 180, 2.1, 180
Liming   0, 324, 0.61, 300, 0.8, 285, 1, 270, 1.55, 250, 2, 180, 2.1, 180

Leaf assimilate fraction (FLVTB)
Guangluai 4   0.00, 0.50, 0.244, 0.50, 0.48, 0.49, 0.66, 0.456, 0.73, 0.327, 0.90, 0.174,
              1.00, 0.00, 2.10, 0.00
Shanyou 6   0.00, 0.54, 0.14, 0.54, 0.292, 0.50, 0.307, 0.48, 0.336, 0.48, 0.545, 0.451,
              0.613, 0.47, 0.67, 0.47, 0.762, 0.17, 0.913, 0.114, 1.00, 0.00, 2.10, 0.00
Shanyou 63   0.00, 0.54, 0.14, 0.54, 0.292, 0.50, 0.307, 0.48, 0.336, 0.48, 0.545, 0.451,
              0.613, 0.47, 0.67, 0.47, 0.762, 0.17, 0.913, 0.114, 1.00, 0.00, 2.10, 0.00
Xiushui 11   0.00, 0.49, 0.50, 0.46, 0.60, 0.42, 0.70, 0.30, 0.80, 0.184, 1.00, 0.00,
              2.10, 0.00
Liming   0.00, 0.49, 0.50, 0.46, 0.60, 0.42, 0.70, 0.30, 0.80, 0.184, 1.00, 0.00,
              2.10, 0.00

Stem assimilate fraction (FSTTB)
Guangluai 4   0.00, 0.50, 0.244, 0.50, 0.48, 0.51, 0.66, 0.544, 0.73, 0.623, 0.90, 0.426,
              1.20, 0.00, 2.10, 0.00
Shanyou 6   0.00, 0.46, 0.14, 0.46, 0.292, 0.50, 0.307, 0.52, 0.336, 0.52, 0.545, 0.549,
              0.613, 0.53, 0.67, 0.53, 0.762, 0.71, 0.913, 0.50, 1.20, 0.00, 2.10, 0.00
Shanyou 63   0.00, 0.46, 0.14, 0.46, 0.292, 0.50, 0.307, 0.52, 0.336, 0.52, 0.545, 0.549,
              0.613, 0.53, 0.67, 0.53, 0.762, 0.71, 0.913, 0.50, 1.20, 0.00, 2.10, 0.00
Xiushui 11   0.00, 0.51, 0.50, 0.54, 0.60, 0.58, 0.70, 0.70, 0.80, 0.616, 1.00, 0.40,
              1.20, 0.00, 2.10, 0.00
Liming   0.00, 0.51, 0.50, 0.54, 0.60, 0.58, 0.70, 0.70, 0.80, 0.616, 1.00, 0.40,
              1.20, 0.00, 2.10, 0.00

Leaf nitrogen concentration (NFLVTB, g N m^{-2} leaf area)
Guangluai 4   0.00, 0.917, 0.20, 1.055, 0.65, 1.501, 1.00, 1.406, 2.00, 0.868, 2.10,
              0.868
Shanyou 6   0.00, 0.917, 0.28, 1.04, 0.362, 1.24, 0.643, 1.53, 0.998, 1.43, 2.10, 1.01
Shanyou 63   0.00, 0.917, 0.28, 1.04, 0.362, 1.24, 0.643, 1.53, 0.998, 1.43, 2.10, 1.01
Xiushui 11   0.00, 0.917, 0.650, 1.428, 0.800, 1.592, 1.00, 1.400, 2.00, 1.100, 2.10,
              1.100
Liming   0.00, 0.917, 0.650, 1.428, 0.800, 1.592, 1.00, 1.400, 2.00, 1.100, 2.10,
              1.100

12.3.2 Model performance

Yields predicted by the ORYZA1 model were compared with reported yields
from nationwide trials of new varieties from various years and sites in the south
of China (Fig. 12.3). Generally, the agreement between simulated and observed
values was good, with the simulated potential yields being higher than the
observed yields in most cases. The few cases where observed yields were higher
than the simulated values could be attributed to differences between the varieties used in the trials and the varieties used in the calibration of the model.

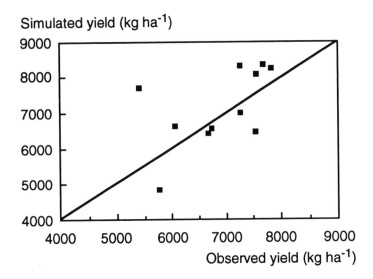

Fig. 12.3. Comparison between experimental yields and those simulated by the crop growth model ORYZA1.

Alternatively, the measured leaf N contents from experimental data that were used as driving data for the model, may not represent the highest leaf N contents possible for each circumstance.

12.4   Effect of Climate Change on Rice Production in China

12.4.1   Input data

The model, calibrated for the five varieties described above, was used to simulate potential rice yields under various scenarios of changed climate, including 'fixed-scenario' changes of CO_2 only (1.5 and 2 times the current level), temperature only (+1°C, +2°C, and +4°C above current temperatures), combinations of these changes, and under scenarios predicted by the GFDL, GISS, and UKMO General Circulation Models (GCMs) described in Chapter 5. In all, 15 different scenarios were simulated, details of which are given in Table 7.5. Representative sites, for which several years of daily weather data were available, were selected from each AEZ, to give a total of ten weather stations. Zone VI was not included in the analysis due the low proportion (0.5%) of rice cultivated there. The selected stations were taken to represent the broad range of different climatic environments in which rice is grown in China. Fuller details of each weather station are listed in Chapters 2 and 7. It was assumed for each scenario that the pattern of day-to-day temperatures within a month is the same as the

Table 12.6. Variety used, Julian date of sowing (SOW) and transplanting (TPLT) of each rice season for each weather station.

Station	AEZ	Early rice			Late rice			Single rice		
		Variety	SOW	TPLT	Variety	SOW	TPLT	Variety	SOW	TPLT
Guangzhou	I	Guangluai	79	109	Shanyou 6	196	221			
Hangzhou	II.1	Guangluai	90	120	Xiushui 11	175	200	Shanyou 63	145	176
Nangjing	II.1							Shanyou 63	126	167
Wuhan	II.1	Guangluai	85	115	Xiushui 11	175	200	Shanyou 63	114	147
Chendu	II.2							Shanyou 63	91	131
Chansha	II.3	Guangluai	90	120	Xiushui 11	176	201			
Fuzhou	II.3	Guangluai	86	116	Xiushui 11	176	201			
Guiyang	III							Shanyou 63	107	147
Beijing	IV							Liming	121	156
Shenyang	V							Liming	121	156

present climate, but that the values are adjusted by the appropriate monthly change predicted by the scenario. A CO_2 level of 680 ppm, double that of the current climate, was used for all three GCM scenarios. Simulations were made using all available years of data for each of the ten weather stations, giving a total of 88 years of weather data. In all, about 2000 simulations were made.

Planting dates used in the simulations were based on those used by farmers practising the different rice cropping systems (Table 12.6). To simulate the effect of the plastic covers used by farmers to protect their young seedlings, the model was modified so that for the first ten days, observed maximum and minimum temperatures were incremented by 5°C and solar radiation was reduced to 70% of its observed value, after which only the observed values for each variable were used. This approach was used for all simulations of early rice, and for single rice at the Beijing and Shenyang sites. The varieties used at each site are shown in Table 12.6.

12.4.2   Effect of fixed increments of temperature and CO_2 on potential yields

When averaged across all sites and all years, there was a general decline in yields for each cropping system as temperature increased from the current

Table 12.7.   Predicted changes (%) in yield for the four temperature increments above current mean temperatures, the three CO_2 levels, and combinations of these, for each of the three planting seasons in China. Values are means of all years available for the relevant sites for each season.

	Temperature increment			
CO_2 level	+0°C	+1°C	+2°C	+4°C
Early rice				
340 ppm	0.00	−7.19	−12.83	−25.42
1.5 × CO_2	24.93	16.34	9.55	−5.89
2 × CO_2	38.95	29.64	22.27	5.30
Late rice				
340 ppm	0.00	−5.58	−11.45	−19.06
1.5 × CO_2	26.23	18.98	11.61	1.42
2 × CO_2	41.11	32.93	24.73	12.98
Single rice				
340 ppm	0.00	−8.79	−17.34	−28.14
1.5 × CO_2	22.05	11.74	1.66	−11.45
2 × CO_2	34.40	23.22	12.32	−2.07

levels (Table 12.7). At current levels of CO_2, for every 1°C increase in temperature, there was a 6.6% decrease in yield from the current values for early rice, 5.2% for late rice, and 8.2% for single-cropped rice. The higher value for single rice reflects the higher temperatures encountered at the time of flowering in this season compared with the other two seasons. Increased temperatures also increased the variability of yields, although to a lesser extent in late rice. At all temperature increments, an increase in the level of atmospheric CO_2 increased yields in each cropping system (Table 12.7), but had little effect on the variability of yields.

Doubling the CO_2 level at current temperatures gave the highest responses in late rice (+41%) and the least in single rice (+34%). A doubled CO_2 concentration was more than able to compensate for the detrimental effects of a +4°C increment for early and late rice, and almost so for single rice. This would suggest that in all but the highest temperature increments predicted by GCMs, the changes in climate produced by a doubling of CO_2 levels is likely to be beneficial to rice production in China.

12.4.3   Effect of predicted GCM scenarios on potential yields

The model simulations indicated that the current potential yields of single-cropped rice were highest, followed by early rice, with late rice the lowest (Table 12.8), agreeing with trends observed in practice. The crop duration of single-cropped rice varieties is usually longer, resulting in higher dry matter production, and the longer grain-filling period makes higher yields possible. During the growth of early rice, temperatures are lower at the start of the season, but higher during the grain-filling stage. Yields in late rice are often lower because of high temperatures in the early stages affecting dry matter production, and increased spikelet sterility caused by cold temperatures during the grain-filling period.

The predicted changes in yields for each scenario and for each rice season are also shown in Table 12.8. Averaged across all seasons and sites, the GFDL, GISS, and UKMO scenarios gave changes of +11.7%, +2.68%, and +6.2% respectively. Averaged across all sites and the three scenarios, there was a +6.6%, +16.3, and −4.5% change predicted for the early, late and single rice, respectively. It seems, therefore, that in a future climate, the yields of early and late rice will increase, but those of single rice will decrease.

In early rice, there were increases in yields predicted for all sites in the GFDL scenario. Under the other two scenarios, small decreases were predicted for two of the sites, but increases for the others. In late rice, increased yields were predicted for all of the sites and for all of the scenarios. This was due to a shortening of the crop duration from the increased temperatures so that the panicle formation and grain-filling periods experienced higher solar radiation regimes than under the current climate. For single-cropped rice, yield changes varied considerably according to the site and the scenario with large increases in yield

Table 12.8. Changes (%) predicted by ORYZA1 in potential rice yield (kg ha^{-1}) in the early rice (a) and late rice (b) in double rice cropping systems and single rice (c) using 2 × CO$_2$ scenarios from the three GCM models.

(a) Early rice

Location	Current (kg ha^{-1})	GFDL % incr.	GISS % incr.	UKMO % incr.	Mean % incr.
Chansha	8,766	9.08	−8.01	5.51	2.19
Fuzhou	8,262	3.01	−6.18	−5.91	−3.03
Guangzhou	6,884	8.44	22.75	12.33	14.51
Hangzhou	9,249	12.11	5.20	−1.84	5.16
Wuhan	9,037	16.92	6.43	18.37	13.91
Mean	8,440	9.91	4.04	5.69	6.55

(b) Late rice

Location	Current (kg ha^{-1})	GFDL % incr.	GISS % incr.	UKMO % incr.	Mean % incr.
Chansha	7,740	26.65	15.04	10.22	17.30
Fuzhou	6,750	28.36	31.04	12.01	23.80
Guangzhou	6,927	12.60	11.16	13.92	12.56
Hangzhou	8,136	15.30	16.78	3.75	11.94
Wuhan	8,505	24.01	15.85	8.37	16.08
Mean	7,612	21.38	17.97	9.65	16.34

(c) Single rice

Location	Current (kg ha^{-1})	GFDL % incr.	GISS % incr.	UKMO % incr.	Mean % incr.
Nangjing	10,114	5.19	8.38	3.38	5.65
Beijing	11,663	1.85	7.15	7.15	5.38
Chendu	9,554	19.93	−67.41	22.74	−8.24
Hangzhou	10,774	2.88	6.34	−9.70	−0.16
Guiyang	11,265	8.28	−11.66	−6.67	−3.35
Wuhan	10,249	−1.01	−39.99	−33.21	−24.74
Shenyang	13,363	−9.63	−0.46	−6.98	−5.69
Mean	10,997	3.93	−13.95	−3.33	−4.45

predicted at Chendu in the GFDL and UKMO scenarios, and large decreases at Chendu and Wuhan in the GISS scenario and at Wuhan in the UKMO scenario.

These large decreases were caused by high temperatures around the time of flowering affecting spikelet fertility. There does, however, appear to be some genotypic variation in this response (Imaki *et al.*, 1987), where 'tolerant' genotypes flower earlier in the day thereby escaping peak temperatures later in the day. Selection by breeders for these genotypes should, therefore, be able to offset this detrimental effect of higher temperatures. An attempt was made to evaluate the effect of selection for more temperature-tolerant genotypes by disabling the routine in the model describing spikelet fertility as a function of temperature and setting it at a constant value of 85% fertility. This value is commonly observed in practice. In this way, yield increases of +7.6%, +6.6% and +6.7% were predicted for the GFDL, GISS, and UKMO scenarios, respectively (Table 12.9), corresponding to the +3.9%, −13.9%, and −3.3% obtained previously (Table 12.8). Only in Shenyang were yield decreases still predicted, indicating that a shortening of the crop duration was the main factor limiting yields at this site.

Table 12.9. Changes (%) predicted by ORYZA1 in potential rice yield (kg ha^{-1}) in the single rice under each of the three GCM models, assuming that spikelet fertility is unaffected by high temperatures.

Location	GFDL % incr.	GISS % incr.	UKMO % incr.	Mean % incr.
Nangjing	5.19	8.38	3.38	5.65
Beijing	1.85	7.15	8.70	5.90
Chendu	19.93	7.91	27.19	18.34
Hangzhou	2.88	8.31	−9.67	0.50
Guiyang	8.28	4.72	3.24	5.42
Wuhan	24.25	10.12	20.26	18.21
Shenyang	−9.00	−0.46	−5.91	−5.13
Mean	7.63	6.59	6.74	6.99

An attempt was made to evaluate the significance of the predicted changes for each season under each GCM scenario in Table 12.8 on the national rice production of China. For this, current national production was stratified according to AEZ and cropping season (Table 12.10). As the largest proportion of China's rice is grown in AEZ II and because more weather stations were available for that AEZ than the others, it was subdivided into its component subzones. Data from Table 12.3 were used to calculate the seasonal production

Table 12.10. Stratification of total annual national production by AEZ and season. AEZ 2 is further stratified into its component subzones (Table 12.3). The weather station(s) representing each AEZ are also shown. Total rice area, mean yields, and proportions of double and single rice (from Table 12.3) are used to calculate seasonal production. Mean yields are assumed to be the same for each season.

AEZ	Weather stations	Area ('000 ha)	Mean yield (t ha^{-1})	% double	% single	Early rice ('000 t)	Late rice ('000 t)	Single rice ('000 t)	Total ('000 t)
I	Guangzhou	6,025	3.96	74	26	17,655	17,655	6,203	41,513
II.1	Hangzhou, Nanjing, Wuhan	9,489	4.77	63	37	28,514	28,514	16,747	73,776
II.2	Chendu	3,187	5.22	7	93	1,165	1,165	15,473	17,802
II.3	Chansa, Fuzhou	10,478	4.23	73	27	32,355	32,355	11,967	76,677
III	Guiyang	2,678	4.34	7	93	813	813	10,796	12,422
IV	Beijing	1,119	4.58	0	100	0	0	5,121	5,121
V	Shenyang	861	5.06	0	100	0	0	4,351	4,351
VI	–	187	4.37	0	100	0	0	818	818
Totals		34,024				80,502	80,502	71,476	232,480

in each AEZ. The predicted changes in yields for each season and zone from Table 12.8 were used, the average being taken where more than one weather station was available for a particular AEZ. As data on average yields for each season were not available (Table 12.3), it was assumed that these were the same for each season, an assumption that may introduce some error into the estimates. These calculated production values were then adjusted by the predicted changes for each season, and summed to give the predicted overall national production (Table 12.11). In this way, changes of +12.2%, +2.0%,

Table 12.11.   Predicted changes in the national rice production of China under the three GCM scenarios.

(a) GFDL

AEZ	Early rice % change	('000 t)	Late rice % change	('000 t)	Single rice % change	('000 t)	Total
I	8.4	19,145	12.6	19,880	0.0	6,203	45,228
II.1	14.5	32,653	19.7	34,119	2.4	17,140	83,913
II.2	0.0	1,165	0.0	1,165	19.9	18,557	20,886
II.3	6.0	34,312	27.5	41,254	0.0	11,967	87,533
III	0.0	813	0.0	813	8.3	11,691	13,316
IV	0.0	0	0.0	0	1.9	5,216	5,216
V	0.0	0	0.0	0	−9.6	3,932	3,932
VI	0.0	0	0.0	0	0.0	818	818
Total		88,088		97,231		75,524	260,842
Overall % change							12.2

(b) GISS

AEZ	Early rice % change	('000 t)	Late rice % change	('000 t)	Single rice % change	('000 t)	Total
I	22.7	21,671	11.2	19,625	0.0	6,203	47,499
II.1	5.8	30,173	16.3	33,166	−8.4	15,336	78,675
II.2	0.0	1,165	0.0	1,165	−67.4	5,043	7,373
II.3	−7.1	30,059	23.0	39,809	0.0	11,967	81,835
III	0.0	813	0.0	813	−11.7	9,537	11,162
IV	0.0	0	0.0	0	7.2	5,487	5,487
V	0.0	0	0.0	0	−0.5	4,331	4,331
VI	0.0	0	0.0	0	0.0	818	818
Total		83,881		94,578		58,722	237,180
Overall % change							2.0

Continued

(c) UKMO

AEZ	Early rice		Late rice		Single rice		Total
	% change	('000 t)	% change	('000 t)	% change	('000 t)	
I	12.3	19,832	13.9	20,112	0.0	6,203	46,147
II.1	8.3	30,871	6.1	30,243	−13.2	14,540	75,654
II.2	0.0	1,165	0.0	1,165	22.7	18,992	21,322
II.3	−0.2	32,291	11.1	35,952	0.0	11,967	80,210
III	0.0	813	0.0	813	−6.7	10,077	11,702
IV	0.0	0	0.0	0	7.2	5,487	5,487
V	0.0	0	0.0	0	−7.0	4,047	4,047
VI	0.0	0	0.0	0	0.0	818	818
Total		84,972		88,285		72,131	245,387
Overall % change							5.6

and +5.6% were predicted under the GFDL, GISS, and UKMO scenarios, respectively. Thus, it would seem that China will benefit from increased rice production under the changed climate of the future.

12.5   Conclusions

The study showed that increased temperatures generally reduced yields, while higher CO_2 levels increased yields, although the degree of change depended both on the GCM scenario used, and on the season in which rice was grown. Overall rice production in China was predicted to change by +12.2%, +2.0%, and +5.6% under the GFDL, UKMO, and GISS scenarios respectively, although those changes varied considerably between seasons. While early and late rice were predicted to have yield increases, single rice in many of the locations was predicted to decline, primarily due to increased temperatures around the time of flowering causing significant spikelet sterility. Selection for varieties that are tolerant to higher temperatures is likely to be able to offset this detrimental effect, so that yield increases may also be obtained for single rice.

Rice Production in The Philippines under Current and Future Climates

13

H.G.S. CENTENO, A.D. BALBAREZ, N.G. FABELLAR,
M.J. KROPFF, AND R.B. MATTHEWS

*International Rice Research Institute, P.O. Box 933, 1099 Manila,
The Philippines*

13.1 Introduction

The Philippines is an archipelago of some 7100 islands located between 4 and 21°N latitude, and 116 and 127°E longitude. The country is bounded by the South China Sea to the west, the Pacific Ocean to the east, the Sulu and Celebes Seas to the south, and the Bashi Channel to the north. The population is around 62 million (1990), with an annual growth rate of 2.2% per year. The population is predominantly rural (57% of the total population), while employment in the agricultural sector accounts for about 44% of the 24 million labor force.

Rice is the main staple for about 75% of the population, constituting about 16% of total crop production and 66% of total cereal production. In recent years, irrigated rice cultivation has increased in intensity, with the extended use of short-maturing varieties allowing up to three crops per year in central Luzon. Problems associated with such intensified systems include yield stagnation (and a possible yield decline) due to factors that are not yet fully understood, deterioration of irrigation systems (given low and decreasing investments in systems maintenance), and increased pests and diseases in some areas. Rainfed lowland rice suffers from uncertain timing of arrival of rains, and drought and submergence – often on the same fields over the course of a single season, or on different fields within a farm over the same season. Weeds, drought, diseases, acidic soils, and soil erosion are major problems of upland rice in The Philippines.

In the short term, few gains are expected in irrigated rice in the Philippines. Modern varieties and corresponding management practices have been adopted in most of the irrigated and favorable rainfed lowland areas.

Deterioration of irrigation schemes and conversion of ricelands to urban uses pose longer-term problems. Some economic gains at the farm level may be expected through further adoption of direct wet-seeded rice and mechanization. Rainfed lowland rice gains may result from some crop intensification in more favorable areas, further adoption of direct dry seeded rice, and development of new rice varieties adapted to rainfed lowland stresses. Upland rice systems in the Philippines are the least sustainable, and future system maintenance will require policies and technologies encouraging a slowing of deforestation rates, the control of soil erosion, and conversion of cereal production systems to mixed systems with perennial crops on sloping lands.

The effect of climate change on rice production in the Philippines is a vitally important issue, particularly as a rapidly expanding population is already encroaching on existing rice-growing lands. Previous simulation studies have been made to evaluate the effect of climate change in the Philippines (Lansigan & Orno, 1991; Lansigan, 1993); these have, however, focused on the effect on yield variability in rainfed rice at a limited number of sites, and have not considered the scenarios predicted by General Circulation Models (GCMs). Escaño & Buendia (1989) used GFDL and GISS $2 \times CO_2$ scenarios for two sites in the Philippines, and predicted declines in crop yields of 15% to 46% for temperature changes only, and increases of 22% and 46% for CO_2 changes only. It is the aim of this study to use weather data from a larger number of locations to simulate the effects of climate change on rice production in the Philippines, both under 'fixed-change' scenarios, where temperature and CO_2 levels are varied in fixed increments, and in scenarios as predicted by three GCMs for a doubled-CO_2 climate.

13.2 Agroecological Zones and Current Rice Production in the Philippines

13.2.1 Agroecological zones

Both the geographic position of the Philippines and their mountainous topography control the seasonal and local variations of the climate. Many of the islands have narrow coastal plains, while the interior is characterized by mountain ranges, generally oriented north–south, and inland plains. The mountain ranges are mostly over 500 m high, with relatively large areas over 1000 m and isolated portions higher than 2000 m. The country frequently suffers from typhoons originating in the Pacific Ocean and moving north-easterly towards Taiwan and Japan. Some 20 typhoons occur every year, mostly during the months of July, August, and September. Although their frequency in October and November is lower, the likelihood of severe damage at this time is higher. Average daily solar radiation reaches a peak of about 25 MJ m^{-2} d^{-1} in April, then gradually drops to values below 15 MJ m^{-2} d^{-1} in December. Maximum and minimum air temperatures follow a similar trend, with highest air temperatures in May. The wet season lasts from June to December, and the dry season from January to May. Annual rainfall varies widely across the country, but in general is higher along the eastern coastal areas which face the Pacific Ocean.

In the agroecological zonation of FAO (see Chapter 7 for details), three

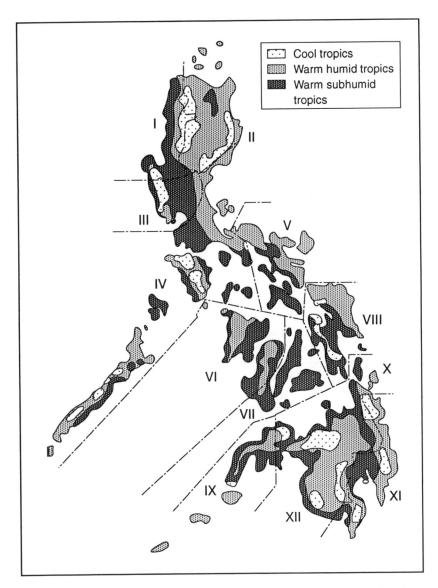

Fig. 13.1. AEZs in the Philippines (according to FAO classification). The administrative regions are also shown (see Table 13.2). (Source: Bureau of Forest Development.)

zones are identified in the Philippines: the warm humid tropics, warm sub-humid tropics and cool tropics (Fig. 13.1). Details of each agroecological zone (AEZ) are given in Table 13.1. The cool tropical areas are in the mountain ranges, while the warm humid areas are in the coastal and low-lying valley areas.

Table 13.1. AEZs of the Philippines.

AEZ	Daily mean temperature (°C)	Length of growing period (d)	Proportion of total land area (%)
Warm humid tropics	> 20	270–365	39.2
Warm subhumid			
tropics	> 20	180–270	48.5
Cool tropics	5–20		12.3
Semi-arid		75–180	
Subhumid		180–270	
Humid		270–365	

Of the approximately 3.4 million ha of ricelands, some 2.1 million ha (61%) are irrigated, 1.2 million ha (35%) are rainfed lowland, and 0.07 million ha (2%) are upland. The annual production of rice, areas harvested, and mean yields for each of the twelve administrative regions in the Philippines (Fig. 13.1) are shown in Table 13.2. Much of the country's irrigated rice is grown on the central plain of Luzon, known as the 'ricebowl' of the Philippines, where average yields during the dry season may be more than 4 t ha^{-1}. The rest comes mainly from various coastal lowland areas and gently rolling erosional plains such as in Mindanao and Iloilo. Rainfed rice, with an average yield of about 2 t ha^{-1}, is found in the Cagayan Valley in northern Luzon, in Iloilo Province, and on the coastal plains of the Visayas region and Ilocos in northern Luzon. Upland rice is grown in both permanent and shifting cultivation systems scattered throughout the archipelago on rolling to steep lands.

13.2.2 The rice crop calendar

Table 13.3 shows the times of planting for each of the main seasons in the Philippines. The largest proportion (approximately 80% of national annual production) of rice is produced in the wet season, with about 18% in the dry season. With the advent of shorter duration varieties, several farmers are able to grow three crops a year or five crops in two years; the contribution of this third season to national annual production, however, is small (about 2%). Time of planting varies slightly from region to region.

Table 13.2. Annual rice production, area harvested, and average yields in 1989 for each administrative region in the Philippines. (Source: Bureau of Agricultural Statistics, 1990).

ID	Region	Total production (10^6 t)	Harvested area (ha)	Average yield (t ha^{-1})
NCR	National Capital Region	152,559	59,330	2.57
I	Ilocos	898,584	312,250	2.88
II	Cagayan Valley	1,033,615	325,290	3.18
III	Central Luzon	1,748,491	499,870	3.50
IV	Southern Tagalog	1,118,085	411,880	2.71
V	Bicol	744,223	295,530	2.52
VI	Western Visayas	1,183,887	456,160	2.60
VII	Central Visayas	207,700	124,380	1.67
VIII	Eastern Visayas	382,954	221,710	1.73
IX	Western Mindanao	399,038	142,780	2.79
X	Northern Mindanao	531,777	170,110	3.13
XI	Southern Mindanao	688,302	205,630	3.35
XII	Central Mindanao	584,047	200,040	2.92
	Total	9,673,262	3,424,960	2.82

Table 13.3. Rice cropping calendar for the 12 administrative regions in the Philippines.

Region	Wet season Transplanting	Wet season Harvest	Dry season Transplanting	Dry season Harvest
I	Jun–Aug	Nov–Dec	Jan–Feb	Apr–May
II	Jun–Aug	Nov–Dec	Jan–Feb	Apr–May
III	Jun–Aug	Nov–Dec	Jan–Feb	Apr–May
IV	Jun–Aug	Oct–Dec	Dec–Jan	Apr–May
V	Jun–Aug	Oct–Dec	Dec–Jan	Apr–May
VI	May–Aug	Oct–Dec	Nov–Dec	Mar–May
VII	May–Aug	Oct–Dec	Nov–Dec	Mar–May
VIII	May–Aug	Sep–Dec	Oct–Dec	Jan–Mar
IX	Jun–Aug	Oct–Dec	Sep–Oct	Jan–Mar
X	Jun–Aug	Nov–Jan	Dec–Feb	Apr–Jun
XI	May–Jul	Sep–Nov	Nov–Dec	Apr–Jun
XII	Aug–Sep	Dec–Jan	Jan–Feb	Jun–Jul

13.2.3 Historical rice production

Large increases in both production and yields have been achieved over the last 25 years (see Fig. 13.2). In 1966, the annual production of rice from

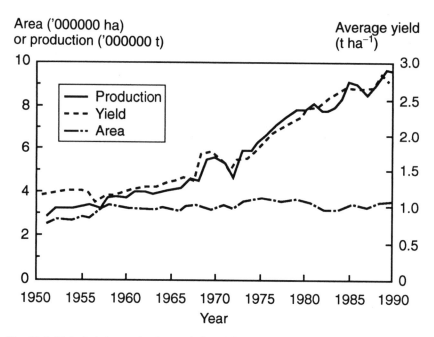

Fig. 13.2. Historical rice production statistics of the Philippines, showing the area sown, average national yields, and annual national production. (Source: IRRI, 1991.)

3.1 million ha was 4.1 million tonnes, giving an average yield of 1.3 t ha^{-1}. By 1991, however, although the area sown to rice had increased by only 10% to about 3.5 million ha, overall production had increased to 9.7 million tonnes, with an average yield of 2.8 t ha^{-1}. Most of this spectacular increase in yield has come from the use of modern varieties, which contributed 87% of the total rice produced in 1991. Despite these increases, however, the Philippines moved from being self-sufficient in rice up until 1983 to having to import 920,000 tonnes by 1991, mainly because of a near-doubling of the population since 1966.

13.3   Model Parameterization and Testing

13.3.1   Parameterization of the crop model for Philippine varieties and conditions

The ORYZA1 potential production rice model, described in Chapter 3 of this book, was used in the study. The model was calibrated from data of the variety IR72 grown in experiments at IRRI, Los Baños. IR72 is a short-maturing cultivar with a total crop duration of 85 Dd[1], similar to IR64 used for the simulations in Chapter 7, and widely grown by farmers in the Philippines. The genotype coefficients and function tables used for IR72 are shown in Table 13.4, and were assumed to be constant for all sites used in the study.

Table 13.4. Genotype coefficients and function tables for the variety IR72 used in the simulations. FLVTB and FSTTB are the functions describing the fraction of new assimilate partitioned to the leaves and stem respectively, and SLATB is the function describing the specific leaf area (cm^2 g^{-1}). All functions are expressed with developmental stage (DVS) as the independent variable.

Genotype coefficient	Name	Units	IR72
Initial relative leaf area growth rate	RGRL	(°Cd)$^{-1}$	0.0077
Stem reserves fraction	FSTR	–	0.3
Basic vegetative period duration	JUDD	Dd	25.4
Photoperiod sensitive phase duration	PIDD	Dd	15.0
Minimum optimum photoperiod	MOPP	h	11.3
Photoperiod sensitivity	PPSE	Dd h^{-1}	0
Duration of panicle formation phase	REDD	Dd	20.1
Duration of grain-filling phase	GFDD	Dd	24.5
Maximum grain weight	WGRMX	mg grain^{-1}	24.9
Spikelet growth factor	SPGF	sp g^{-1}	64.9

FLVTB = 0.000, 0.539, 0.078, 0.545, 0.245, 0.559, 0.491, 0.548, 0.718, 0.465, 0.893, 0.152, 1.229, 0.128, 1.729, 0.000, 2.100, 0.000

FSTTB = 0.000, 0.461, 0.078, 0.455, 0.245, 0.441, 0.491, 0.452, 0.718, 0.535, 0.893, 0.462, 1.229, 0.000, 1.729, 0.000, 2.100, 0.000

SLATB = 0.000, 649, 0.157, 500, 0.333, 333, 0.650, 279, 0.787, 210, 1.000, 194, 1.458, 170, 2.000, 171, 2.100, 171

13.3.2 Model performance

Testing of the model for Philippine conditions has already been described in Chapter 3 (Section 3.14), using data from IR72 grown in potential production experiments at IRRI, Los Baños. Good agreement was obtained between simulated and observed values of growth, total dry matter production, and yield for two seasons (Fig. 3.10). Similarly, when the model was evaluated using experimental data from a wide range of treatments for IR58 and IR64 from several years, relative yield differences between experiments were adequately

---

[1] Dd represents a 'developmental day'. See Chapter 3 of this book for further details.

simulated (Fig. 3.11), although for IR64, yields were consistently overestimated. Nevertheless, as the model could adequately simulate trends, we considered it to be able to predict relative changes in yields brought about by changes in climate.

13.4   Effect of Climate Change on Rice Production in the Philippines

13.4.1   Input data

The ORYZA1 model, calibrated for the variety IR72 as described above, was used to simulate potential rice yields under various scenarios of changed climate, including 'fixed-scenario' changes of CO_2 only (1.5 and 2 times the current level), temperature only (+1°C, +2°C, and +4°C above current temperatures), combinations of these changes, and under scenarios predicted by the GFDL, GISS, and UKMO GCMs described in Chapter 5. In all, 15 different scenarios were simulated, details of which are given in Table 7.5. For each scenario, it was assumed that only the absolute values of solar radiation and temperature were altered by the change in climate, but that the annual pattern of variation remained the same. Fifteen locations covering the major rice-growing regions of the Philippines and for which daily weather data was available were selected. Extra stations were included in the analysis for this chapter than were available at the time the simulations for Chapter 7 were made. Details of the stations used are shown in Table 13.5 and Table 7.3. For each site, dates of sowing and transplanting were determined from Table 13.3, and were the same as used for the simulations in Chapter 7. Transplanting was assumed to take place at 20 days after sowing (DAS) in each case.

13.4.2   Effect of fixed increments of temperature and CO_2 on potential yields

The relative effect on potential yields of varying temperature and CO_2 in fixed increments individually and in combination is shown in Table 13.6. At current temperatures, with no increment, a doubling in the level of CO_2 resulted in a yield increase of about 30%, which was also remarkably consistent for the three seasons. This value is lower than that predicted for other country studies in this report, but is more in agreement with the 30% reported by Kimball (1983). At current levels of CO_2, increasing the temperature always decreased yields, but much more so in the dry season due to the already high temperatures prevalent during that period. Decreases were less marked in the transition season. Average decreases in yield per 1°C increase were −7.2%, −10.3%, and −4.4% $(°C)^{-1}$ for the wet, dry and transition seasons, respectively.

Table 13.5. Details of the weather stations used in the study.

ID no.	Location	Region	Longitude (°E)	Latitude (°N)	Elevation (m)	Mean ann. temp. (°C)	Mean ann. solar rad. (MJ m^{-2} d^{-1})
1	MMSU	I	124.30	18.00	854	27.0	18.4
2	Banaue	II	121.07	16.92	1040	21.3	15.1
3	Solana	II	121.68	17.65	21	27.1	18.5
4	CLSU	III	120.90	15.72	76	27.1	18.9
5	Guimba	III	120.78	15.65	66	27.3	21.0
6	H. Luisita	III	120.60	15.43	32	27.1	18.3
7	Munoz	III	120.93	15.75	48	27.2	20.8
8	Cavinti	IV	121.50	14.28	305	24.4	16.4
9	IRRI Wetland	IV	121.25	14.18	21	26.8	16.1
10	PNAC	IV	118.55	9.43	7	27.2	18.4
11	UPLB	IV	121.25	14.17	21	27.1	16.9
12	Albay	V	123.70	13.38	–	26.7	16.2
13	La Granja	VI	122.93	10.40	84	27.1	13.7
14	VES	VI	122.58	10.77	14	28.5	18.3
15	Betinan	IX	123.45	7.90	45	26.6	16.0
16	Butuan	X	125.73	11.03	60	27.3	16.6

Table 13.6. Yield changes (%) predicted for the wet, dry, and transition planting seasons under fixed scenarios of increased CO_2 level and temperature increments of +1°C, +2°C, and +4°C. Values are overall means of the 15 weather stations and all years available for each.

	Temperature increment			
	+0°C	+1°C	+2°C	+4°C
Wet season				
340 ppm	0	−6.54	−13.81	−32.17
1.5 × CO_2	18.08	10.54	1.93	−20.37
2 × CO_2	28.12	20.05	10.69	−13.78
Dry season				
340 ppm	0	−7.83	−19.79	−52.95
1.5 × CO_2	19.12	9.70	−4.77	−44.24
2 × CO_2	29.70	19.37	3.53	−39.45
Transition season				
340 ppm	0	−4.48	−8.45	−18.10
1.5 × CO_2	20.50	15.25	10.48	−1.79
2 × CO_2	31.95	26.31	21.16	7.35

In the wet and dry seasons, an increment of +4°C always produced a decrease in yields at all increments of CO_2; in the transition season, however, doubling the CO_2 level was able to compensate for the detrimental effect of this increase in temperature, so that a small increase was predicted.

13.4.3   Effect of predicted GCM scenarios on potential yields

The simulated yields for the current climate and for the three GCM scenarios in the wet, dry and transition seasons are shown in Tables 13.7, 13.8 and 13.9, respectively. Current wet-season potential yields ranged from 8.3 t ha^{-1} at La Granja to more than 13.8 t ha^{-1} at Banaue. Similar trends were also simulated for the dry and transition seasons. La Granja, in the Western Visayas, is characterized by low annual solar radiation receipts, resulting in lower crop growth rates, while Banaue, in northern Luzon, is situated at more than 1000 m altitude and experiences low mean annual temperatures (Table 13.5), so that crop durations are longer.

Table 13.7.   Changes predicted by ORYZA1 in potential rice yields (kg ha^{-1}) at each weather station during the wet season, using $2 \times CO_2$ scenarios from the three GCM models.

Location	Current kg ha^{-1}	GFDL kg ha^{-1}	GFDL % change	GISS kg ha^{-1}	GISS % change	UKMO kg ha^{-1}	UKMO % change
Banaue	13,888	14,019	0.9	13,490	−2.9	14,810	6.6
Solana	10,987	9,662	−12.1	7,619	−30.7	9,349	−14.9
Guimba	10,754	11,418	6.2	9,382	−12.8	12,342	14.8
Munoz	11,324	12,081	6.7	12,089	6.8	13,479	19.0
CLSU	10,835	11,098	2.4	10,797	−0.4	12,139	12.0
IRRI Wetland	8,968	9,791	9.2	8,956	−0.1	9,771	9.0
Cavinti	10,517	10,927	3.9	10,459	−0.6	11,038	5.0
PNAC	10,250	12,057	17.6	10,956	6.9	11,814	15.3
UPLB	9,360	9,945	6.3	7,149	−23.6	8,977	−4.1
Albay	10,013	10,447	4.3	6,218	−37.9	8,369	−16.4
VES	10,439	12,280	17.6	11,330	8.5	12,110	16.0
La Granja	8,370	9,478	13.2	7,098	−15.2	8,307	−0.8
Betinan	10,688	12,753	19.3	11,691	9.4	11,923	11.6
Butuan	10,480	12,030	14.8	8,706	−16.9	7,243	−30.9
MMSU	10,156	12,044	18.6	9,303	−8.4	10,664	5.0
Average	10,469	11,335	8.3	9,683	−7.5	10,822	3.4

Table 13.8. Changes predicted by ORYZA1 in potential rice yields (kg ha⁻¹) at each weather station during the dry season, using 2 × CO_2 scenarios from the three GCM models.

Location	Current kg ha⁻¹	GFDL kg ha⁻¹	GFDL % change	GISS kg ha⁻¹	GISS % change	UKMO kg ha⁻¹	UKMO % change
Banaue	14,401	16,612	15.4	15,462	7.4	18,207	26.4
Solana	11,459	9,536	−16.8	5,171	−54.9	14,283	24.6
Guimba	12,676	9,154	−27.8	3,694	−70.9	14,583	15.0
Munoz	13,471	15,426	14.5	11,088	−17.7	18,149	34.7
CLSU	13,364	11,771	−11.9	6,681	−50.0	16,456	23.1
IRRI Wetland	12,325	14,816	20.2	11,869	−3.7	8,255	−33.0
Cavinti	14,128	17,262	22.2	15,883	12.4	14,774	4.6
PNAC	13,140	16,057	22.2	11,703	−10.9	14,958	13.8
UPLB	12,497	11,179	−10.5	6,213	−50.3	3,021	−75.8
Albay	12,192	13,260	8.8	10,611	−13.0	7,669	−37.1
VES	11,866	11,895	0.2	7,076	−40.4	4,043	−65.9
La Granja	8,851	8,190	−7.5	4,605	−48.0	2,437	−72.5
Betinan	10,769	12,332	14.5	9,586	−11.0	11,700	8.6
Butuan	12,278	11,466	−6.6	6,809	−44.5	2,879	−76.6
MMSU	12,500	11,715	−6.3	6,397	−48.8	9,388	−24.9
Average	12,394	12,711	2.6	8,857	−28.5	10,720	−13.5

Predicted changes in yield ranged from a −76% decrease at Butuan in the dry season, to a massive +256% increase at Banaue in the transition season, both in the UKMO scenario. This huge increase in Banaue was due to simulated current rice yields in the transition season being limited by low temperatures caused by its high altitude; the warmer temperatures of a changed climate allowed rice to be grown in this region where it would not have been economic or possible before. However, as three crops of rice are not grown at present at this location anyway, these large increases are not significant. When averaged over all sites, the GISS scenario was the most detrimental in both the wet and dry seasons, and the GFDL the least severe, reflecting the predicted increments in temperature for each scenario.

Averaged over all sites and all seasons, and using a weighting of 80 : 18 : 2 for the relative contributions to overall production from each season (Section 13.2.2), the predicted changes in yield were +7.3%, −11.5%, and +0.0% for the GFDL, GISS, and UKMO scenarios, respectively (Table 13.10). These values compare with the +14.1%, −11.8%, and −4.7% changes predicted in Chapter 7 (Table 7.9) for the Philippines; the ranking of the GCM scenarios is the same, and differences in the absolute changes can be ascribed to the different variety and the extra weather stations used in this analysis.

Table 13.9. Changes predicted by ORYZA1 in potential rice yields (kg ha^{-1}) at each weather station during the transition planting season, using 2 × CO_2 scenarios from the three GCM models.

Location	Current kg ha^{-1}	GFDL		GISS		UKMO	
		kg ha^{-1}	% change	kg ha^{-1}	% change	kg ha^{-1}	% change
Banaue[1]	3,467	11,470	230.8	12,365	256.6	12,357	256.4
Solana	10,144	11,592	14.3	11,138	9.8	10,194	0.5
Guimba	11,995	13,929	16.1	11,473	−4.4	10,867	−9.4
Munoz	11,333	13,906	22.7	13,360	17.9	12,389	9.3
CLSU	9,994	11,906	19.1	11,742	17.5	10,393	4.0
IRRI Wetland	8,261	9,206	11.4	8,611	4.2	8,670	5.0
Cavinti	11,511	11,386	−1.1	10,008	−13.1	9,719	−15.6
PNAC	9,808	11,639	18.7	11,123	13.4	11,374	16.0
UPLB	8,175	9,447	15.6	8,837	8.1	8,738	6.9
Albay	7,731	8,692	12.4	8,138	5.3	8,053	4.2
VES	9,845	12,111	23.0	11,300	14.8	11,129	13.0
La Granja	7,652	9,425	23.2	8,210	7.3	7,612	−0.5
Betinan	9,981	12,423	24.5	10,747	7.7	11,302	13.2
Butuan	8,625	10,708	24.2	9,969	15.6	9,935	15.2
MMSU	11,773	13,681	16.2	11,087	−5.8	11,968	1.7
Average	9,353	11,435	22.3	10,541	12.7	10,313	10.3

[1] A crop is not normally grown at Banaue in the transition period as low temperatures mean there is insufficient time to grow three crops in a year. Data are nevertheless shown to indicate that a warmer climate can result in comparable yields to other sites.

The predicted changes in rice yields can be used to estimate the changes in overall rice production in the Philippines under each GCM scenario, stratifying total production to weight the predicted changes to account for uneven distribution of production over the country. This was done by grouping the weather stations into their respective administrative regions (Table 13.2), and calculating the mean predicted change for each group. The mean change for each region was then used to adjust the current rice production for that region (Table 13.2). Administrative regions, rather than AEZs, were used for stratification as rice production data were not available for each AEZ. Results of this analysis are shown in Table 13.11.

The predicted changes in national rice production were +6.6%, −14.0%, and +1.1% for the GFDL, GISS, and UKMO scenarios, respectively. These compare closely with the mean changes of +7.3%, −11.5%, and +0.0% calculated previously without stratification. Clearly, the predicted changes depend on the GCM scenario used, and until a closer agreement between GCM predictions of

Table 13.10. Changes predicted by ORYZA1 in potential annual mean rice yields (kg ha^{-1}) at each weather station, using 2 × CO_2 scenarios from the three GCM models. Yields are the averages of the wet, dry, and third planting season values from Tables 13.7, 13.8, and 13.9, weighted in the proportion 80 : 18 : 2 to account for the contribution to total annual production for each season.

Location	Current kg ha^{-1}	GFDL kg ha^{-1}	GFDL % change	GISS kg ha^{-1}	GISS % change	UKMO kg ha^{-1}	UKMO % change
Banaue	13,772	14,435	4.8	13,822	0.4	15,372	11.6
Solana	11,055	9,678	−12.5	7,249	−34.4	10,254	−7.2
Guimba	11,125	11,061	−0.6	8,400	−24.5	12,716	14.3
Munoz	11,711	12,720	8.6	11,934	1.9	14,298	22.1
CLSU	11,273	11,235	−0.3	10,075	−10.6	12,881	14.3
IRRI Wetland	9,558	10,684	11.8	9,473	−0.9	9,476	−0.9
Cavinti	11,187	12,076	8.0	11,426	2.1	11,684	4.4
PNAC	10,761	12,769	18.7	11,094	3.1	12,371	15.0
UPLB	9,901	10,157	2.6	7,014	−29.2	7,900	−20.2
Albay	10,360	10,918	5.4	7,047	−32.0	8,237	−20.5
VES	10,684	12,207	14.3	10,564	−1.1	10,638	−0.4
La Granja	8,442	9,245	9.5	6,672	−21.0	7,237	−14.3
Betinan	10,688	12,671	18.5	11,293	5.7	11,870	11.1
Butuan	10,767	11,902	10.5	8,390	−22.1	6,511	−39.5
MMSU	10,610	12,018	13.3	8,816	−16.9	10,460	−1.4
Average	10,793	11,585	7.3	9,551	−11.5	10,794	0.0

the likely changes in climate at the local level is reached, predicted effects on crop yields and production cannot be more accurate.

As pointed out in Chapter 7 of this book, the main reason for differences in predicted changes between GCM scenarios was due to the sensitivity of spikelet fertility to changes in daily maximum temperature in the range 34–36°C (Fig. 4.6). A small increase in temperature in this range can lead to large decreases in spikelet fertility, changing an increase in yield rapidly into a significant decrease. Many of the locations in the study already experience temperatures around the time of flowering near this range in the current climate; small differences in the temperature increases predicted by each scenario can therefore give a large range of responses. It seems likely that varieties with a greater tolerance of spikelet fertility to temperature can be selected for; the current range of variation observed between genotypes (Satake & Yoshida, 1978) seems to be adequate enough to offset the detrimental effect of the increase in temperatures predicted by the GCM scenarios (see Fig. 7.10). As the changes are predicted to occur gradually over the next century, there should be sufficient time for plant breeding programs to select appropriate varieties.

Table 13.11.   Estimated changes in rice production from each of the administrative regions in the Philippines, and from the whole country. Proportional changes are the averages of those predicted for all stations (Table 13.10) in each of the administrative regions.

Region	Current t	GFDL		GISS		UKMO	
		% change	t	% change	t	% change	t
NCR	152,559	2.6	156,476	−11.1	135,669	16.9	178,319
I	898,584	−3.8	864,238	−17.0	745,538	2.2	918,241
II	1,033,615	−3.8	994,108	−17.0	857,571	2.2	1,056,226
III	1,748,491	2.6	1,793,379	−11.1	1,554,911	16.9	2,043,730
IV	1,118,085	10.2	1,232,604	−6.2	1,048,730	−0.4	1,113,437
V	744,223	5.4	784,357	−32.0	506,260	−20.5	591,716
VI	1,183,887	11.9	1,324,583	−11.1	1,053,064	−7.4	1,096,816
VII	207,700	11.9	232,384	−11.1	184,749	−7.4	192,424
VIII	382,954	11.9	428,465	−11.1	340,637	−7.4	354,789
IX	399,038	18.5	473,040	5.7	421,617	11.1	443,166
X	531,777	10.5	587,861	−22.1	414,386	−39.5	321,605
XI	688,302	13.3	779,593	−16.9	571,880	−1.4	678,580
XII	584,047	13.3	661,510	−16.9	485,259	−1.4	575,798
Totals	9,673,262		10,312,598		8,320,271		9,564,847
% change from current			6.6		−14.0		−1.1

13.5   Conclusions

Simulation of the likely effects of a doubled-CO_2 climate on rice production in the Philippines predicted increases in annual production for most regions under the GFDL scenario, and decreases for most regions under the GISS scenario. Both increases, mostly in the north of the country, and decreases, mostly in the south, were predicted for yields under the UKMO scenario. Overall, it was predicted that the national rice production of the Philippines would change by +6.6%, −14.0%, and +1.1% for the GFDL, GISS, and UKMO doubled-CO_2 scenarios respectively. Most of these changes reflect the effect of the scenarios on rice production in the wet season which contributes 80% towards current annual production. Changes predicted in the dry and transition seasons, while dramatic in some cases (e.g. +250% increases at Banaue), were not so significant due to their lesser contribution to overall production. Nevertheless, a warmer climate may allow a third rice crop to be grown in these cooler regions, thereby altering the proportion of rice produced in each season.

The decline under the GISS scenario was due to the higher temperatures predicted, which reduced spikelet fertility significantly. Selection for genotypes with higher tolerance of spikelet fertility to temperature should offset any detrimental effect of a changed climate.

THE IMPACT OF GLOBAL CLIMATE CHANGE ON RICE PRODUCTION IN ASIA: CONCLUSIONS OF THE STUDY

III

Summary and Limitations   14

R.B. Matthews[1]

Research Institute for Agrobiology and Soil Fertility, Bornsesteeg 65, 6700 AA Wageningen, The Netherlands

14.1 Overview

In recent years, the prospect of changes in the earth's climate has stimulated considerable research interest in attempting to predict how production of crops will be affected. The effect of changes in climate on rice production is of particular interest, not only because of its importance in the diet of millions of people, but also because recent intensification of rice production, particularly in the Asian region, has itself contributed to global warming through the release of methane into the atmosphere. The purpose of the work described in this book is to provide estimates of the likely effect of climate change on rice production in the Asian region. Although the influence of rice production on methane evolution is outside the scope of the book, the work should provide a baseline for evaluating this relationship in more detail in subsequent studies.

    Climate change may affect overall rice production by altering both the area of production and the yield per unit area. Areal extent is influenced by changes in the rice-growing environment, either by allowing rice to be grown where or when it is not grown currently, or by reducing yields to below a threshold where it is not economic for the farmer to grow rice. In both cases, the economics of input supply, marketing infrastructure, and financial return play a major role; whether crops will or will not be grown in a given area depend on the farmers' perception of whether it is financially viable to do so. Yields per unit area are influenced by the direct effect of climate change on the physiological mechanisms of the rice growing plant. In general, an increase in

---

[1] Seconded to International Rice Research Institute, P.O. Box 933, 1099 Manila, The Philippines.

CO_2 level increases growth and yield, mainly through its effect on the crop's photosynthetic processes. An increase in temperature, on the other hand, generally decreases yield by speeding up crop development so that it matures sooner, so that there is not as long a period in which to produce a yield. In addition, in rice, both high and low temperatures may affect panicle sterility, which may limit yields even if there is sufficient growth in other plant components. It is often difficult to predict the overall effect of an increase in CO_2 and temperature, as the result depends on the relative effects of both variables in the particular combination occurring for a given region. Simulation models offer a way in which this can be done.

In this study, we addressed these issues using simulation and systems analysis methods at three different levels. Firstly, we used Geographical Information System (GIS) data to estimate the effect of climate change on changes in the area and distribution of rice-growing areas (Chapter 6). Secondly, we used two different crop simulation models, calibrated for the two main rice ecotypes, *indica* and *japonica*, to predict changes in rice production at the regional level (Chapter 7), and thirdly, we used the same two crop models, calibrated for local varieties and conditions in a number of Asian countries, to evaluate likely changes in a more detailed analysis (Chapter 8). For the crop simulations, we first investigated the effect of various levels of CO_2, since the increase in CO_2 level is the most certain and virtually uniform change predicted. Next, we simulated the effects of fixed increments of temperature, along with combinations of CO_2 and temperature. Finally, as they are the least certain, the effect of the changes in temperature and solar radiation predicted by three General Circulation Models (GCMs) were simulated. In this way, the simulations are ordered according to the certainty of the environmental change being considered.

14.2   General Findings

The area under rice cultivation may be increased by higher global temperatures allowing extra rice crops to be grown in the cycle, either by extending the current growing season in areas where the growing season is restricted, or by shortening the duration of current varieties. This may allow expansion of rice cultivation in northern latitudes where cold temperatures render its current development impossible. Economic incentives would greatly enhance this expansion, but would only be possible if soils and management practices in these new areas were suitable, and population density and agricultural infrastructure were also adequate. However, this expansion may be at the expense of areas closer to the equator, where increased temperatures above already high values may cause increased incidence of spikelet sterility, or where water reserves are not sufficient to meet the increased evaporative demand.

In this study, an attempt was made to classify the rice-growing regions of Asia into agroclimatic zones based on accumulated day degrees and precipitation (Chapter 6). The changes in areal extent of these zones under the GFDL, GISS, and UKMO GCM scenarios (see Chapter 5) were then compared. The results indicated that there would be a shift northwards by many of the zones, but losses in area of zones currently in the equatorial latitudes were compensated by gains in zones at higher latitudes. In particular, large increases in the areas of irrigated rice ecosystems were predicted to occur in eastern India, Bangladesh, southern and eastern China and several of the south-east Asian countries. However, while large changes in the areal extent of individual zones were predicted, overall rice cultivation would not be greatly affected. This is similar to the results of the study of Solomon & Leemans (1990), who concluded that rice cultivation would only lose 1.3% of its current extent. However, since the Asian population continues to increase significantly, even this small reduction in cultivation area could have significant economic impacts.

While an evaluation of climate change effects on areal distribution of rice-growing regions can be made by considering environmental parameters only, the effect on rice yields throughout the region requires the crop and its physiology to be also taken into account. In recent years, a number of controlled-environment studies have increased our understanding of the effect of increased temperature and CO_2 on crop growth and development. The use of crop simulation models offers a way in which this experimental information can be extrapolated both in time and space. These models describe mathematically the processes involved at the crop level, and sometimes even those at the single leaf level, and, together with GCMs, enable a synthesis of information from all of the relevant levels of organization. However, the models available at present, both GCMs and crop models, are far from perfect, and often contain assumptions that are not fully tested. Some uncertainty, therefore, is inherent in any predictions by the models. To gain some idea of the degree of this uncertainty, we have used predictions by the three GCMs, GFDL, GISS, and UKMO, and two crop simulation models (ORYZA1 and SIMRIW). While it is not known how close the predictions of these models of future effects are to reality, this approach of using several models can, at least, indicate the sensitivity of the rice production system to the various assumptions made in the models, thereby providing a better understanding of the underlying processes involved, and indicate the general response of the system to changes in climate.

We chose to use potential production models based on the assumption that the proportional changes in yield brought about by climate change at this level are similar to those at other levels of production, such as when water or nitrogen is limiting, or if pests or diseases are present. The two models differ mainly in the way dry matter production and partitioning, leaf area development and phenological development, are calculated (Section 7.4.1). ORYZA1 calculates dry matter production as a function of light, CO_2, and temperature, by considering photosynthetic processes at the leaf level and integrating these over the canopy to obtain crop level values. Respiration is also modeled explicitly as a function of temperature, and partitioning of dry matter is according to phenology-dependent functions. We reasoned that it was useful to include a model containing detail of processes at the leaf level, as much of the knowledge of CO_2 and temperature effects on growth processes is available at this level of organization. The SIMRIW model, on the other hand, uses a simpler radiation-use-efficiency (RUE) relationship between intercepted solar radiation and growth, in which respiration is implicit. CO_2 effects are accounted for using a curvilinear function relating RUE to CO_2 concentration. Temperature is assumed not to affect RUE.

Leaf area development is calculated in ORYZA1 using a temperature-dependent relationship in the early stages, and a growth-related relationship later. SIMRIW decouples growth and leaf area development completely, using a temperature-dependent relationship throughout the period until flowering. Only partitioning of dry matter to the grains at final harvest is described, and that to leaves and stems is not considered at all. Leaf senescence in both models is handled in similar ways, as an empirical function of phenological stage. Phenology routines in both models are similar for ambient temperatures up to about 30°C, but ORYZA1 assumes that the rate of development decreases at temperatures in excess of this value, while SIMRIW assumes that there is no further change.

Both of the models operate on a daily time step, and require values of solar radiation, and minimum and maximum temperatures as input. This data was collected from various rice-growing areas in Asia, giving a total of 68 weather stations. For each station, up to 30 years of data were available. Various procedures were followed to validate the data to ensure that errors in measurement and recording did not cause the models to give incorrect simulations (Chapter 2).

In general, there was good agreement between the two crop models (Chapter 7), although they tended to deviate at both the high and low ends of the yield range. At the low end of the range, SIMRIW predicted higher yields than ORYZA1, which was due to the difference in the way respiration is accounted for, as described above. At the high end of the yield range, there was a slight tendency for SIMRIW to predict lower potential yields than ORYZA1, which was due to the relationship between CO_2 level and crop growth rate being lower at the 680 ppm level in SIMRIW. Both models predicted that at all CO_2 levels, an increase in temperature would cause a decline in yields, and at all temperature increments, an increase in CO_2 would increase yields, similar to the findings of other studies (e.g. Bachelet et al., 1993). Averaged over all sites and years, ORYZA1 predicted a −7.4% change in yields for every 1°C increase in temperature at the current level of CO_2, while SIMRIW predicted −5.3% °C^{-1}, the difference, again, being due to the respiration rate calculations. Both of these values agree closely with the 7 to 8% °C^{-1} decrease measured in controlled-environment experiments (Baker et al., 1992b).

Using the GCM scenarios, and weighting the prediction values by current annual production in each country and by the contributions from the planting seasons at each site, the ORYZA1 model predicted changes in overall regional annual rice production of +6.5%, −4.4%, and −5.6% for the GFDL, GISS, and UKMO scenarios respectively. The corresponding changes predicted by SIMRIW were +4.2%, −10.4%, and −12.8% (Tables 7.9 and 7.10). The ranking in each case corresponds to the temperature increase predicted by each GCM (Table 5.1), and is similar to that found in the study of Rosenzweig *et al.* (1993). Taking the average of all of these estimates, it would appear that rice production in the Asian region may decline by −3.8% under the climates of the next century. However, for both crop models, the predicted changes varied considerably between countries, with declines in yield predicted under the GISS and UKMO scenarios for Thailand, Bangladesh, and western India, while increases were predicted for Indonesia, Malaysia, and Taiwan, and parts of India and China. Such changes are likely to have a significant effect on future trading relationships within the south-east and eastern Asian region.

The detailed country studies provide interesting insights into the effects of climate change at the national level. In Japan, it was predicted that yields will substantially increase in northern and north-central parts of the country, but decline in south-central and south-western regions, so that national rice production would not change significantly from the current level (Chapter 8). Variability of yields is also likely to increase, reflecting increased high temperature-induced spikelet sterility in the southern areas.

In India, the annual national rice production was predicted to increase under the GCM scenarios used (Chapter 9). This was mainly due to an increase in yields of main season crops where the fertilizing effect of the increased CO_2 level is more than able to compensate the crop for any detrimental effects of increased temperatures. Although large decreases were predicted for second season crops at many of the locations due to high temperatures being encountered, the relatively low proportion of total rice produced in this season meant that its overall effect on national production was small.

The results from Malaysia indicated that national rice production is likely to increase significantly by 20–30% in all the scenarios considered (Chapter 10). This was because temperatures during the growing seasons in the current climate are about 26°C, which, even with the temperature increments predicted by the GCMs, do not rise to a level where spikelet fertility is likely to be influenced by high temperatures. The increased temperatures also did not shorten the duration of the crop sufficiently to negate the beneficial effect of the increased CO_2 level.

In South Korea, changes of −9%, 1%, and −25% in annual national production were predicted for the GFDL, GISS, and UKMO scenarios respectively (Chapter 11). Overall, therefore, it seems that a decline in production may occur. Detailed analysis showed that this decline was mainly due to a decrease in yields in the lower altitude areas, where temperatures at the time of flowering were high, resulting in spikelet sterility. The effect of increased temperature was less marked at higher altitudes.

Annual rice production in China was predicted to change by +12.2%, +2.0%, and +5.6% under the GFDL, UKMO, and GISS scenarios respectively, although those changes varied considerably between seasons (Chapter 12). 'Early' and 'late' rice were predicted to have yield increases, but 'single' rice in many of the locations was predicted to decline, primarily due to significant spikelet sterility. Rice planted in the early and late seasons was able to avoid high temperatures around the time of flowering.

In the Philippines, national rice production was predicted to change by +6.6%, −14.0%, and +1.1% for the GFDL, GISS, and UKMO scenarios (Chapter 13). The average of these would indicate a decline of −2.1%. These changes were mainly due to the effect on wet season production, which contributes 80% towards current annual production. Changes predicted in the dry and transition seasons, while spectacular in some cases, were not so significant due to their lesser contribution to overall production. Nevertheless, a warmer climate may allow a third rice crop to be grown in regions where only two crops can be grown at present.

There was good agreement in the predicted changes in production for each country between the regional analysis of Chapter 7, and the individual country studies in Chapters 8–13 (Fig. 14.1). The outlying points are for India, and to a lesser extent, China. In India, these discrepancies were due to the difference in spikelet formation efficiency between the varieties used; in Chapter 7, IR64 was used with an SPGF value of 49.2 spikelets g^{-1} (Table 7.1), while in Chapter 9, IR36 was used with a SPGF value of 64.9 spikelets g^{-1} (Table 9.3). The higher number of spikelets formed in IR36 meant that it could sustain higher numbers of infertile spikelets caused by high temperatures than could IR64, without affecting the yield overly much. As both values of SPGF were derived from experimental data, the difference between these varieties highlights the importance of the role that the selection for appropriate genotypes will have in offsetting any detrimental effects of climate change in the future.

For both China and India, differences in the stratification method used for estimating national production were also partly responsible for the discrepancy between the regional analysis and the individual country analysis. In China, both the FAO zones and national agroecological zoning schemes were used, whereas in India, FAO zones and administrative regions were used. Without an independent evaluation, it is difficult to say which approach of stratification is the more realistic, although intuition would suggest that the more detailed national agroecological zoning is superior to the lower resolution FAO zoning, but that FAO zoning is more accurate than administrative zoning. These differences highlight the need to obtain more weather data from a greater number of weather stations in these two countries so that the coverage is more representative of their rice-growing areas. The good agreement for the other countries reflects the greater homogeneity in their rice-growing environments. It is also interesting to note that, based on the analysis of the individual country

Change in national production
in individual country analysis (%)

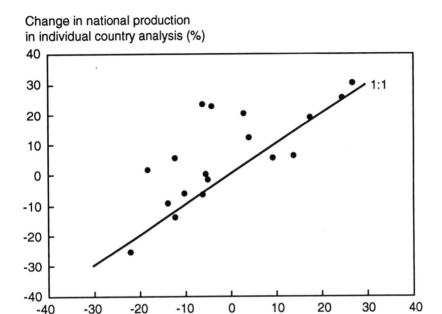

Change in national production in regional analysis (%)

Fig. 14.1. A comparison of the predicted yield increases calculated in the regional analysis using the two varieties IR72 and Ishikawi (Chapter 7) and those calculated in the individual country studies using local varieties (Chapters 8–13). Planting dates and climate data were the same for each comparison.

chapters, the greatest yield increases were predicted for the tropical countries of Malaysia and India, whereas decreases were predicted for the more temperate countries of South Korea and Japan. This contrasts with the conclusions of Rosenzweig *et al.* (1993), who predicted that crop yields are likely to decline in the low-latitude regions as they are currently grown nearer their limits of temperature tolerance, but could increase in the mid- and high-latitudes, where increased warming will extend growing seasons. However, in most tropical countries, the main rice-growing season is during the monsoon, when temperatures are lower than at other times of the year, whereas in the more temperate regions the main season is in the summer, when temperatures, particularly around the time of flowering, can be quite high. This point emphasizes the need to complement large-scale global and regional studies with more detailed studies at the national, and even district, level.

The response of spikelet sterility to temperature emerged as a major factor determining the differential predictions of the effects of climate change on each scenario (e.g. Table 7.11). For many rice varieties, maximum daily temperatures in excess of about 33–34°C at the time of flowering cause severe damage to the spikelets, resulting in significant numbers becoming sterile and not able to form grains. Where current mean temperatures are already high (in the region of 30°C), a further increase predicted for the scenario raises the maximum temperatures encountered to this sensitive range. Fortunately, many of the extreme effects on rice yield were predicted for the second growing season where temperatures are higher, but contribution to overall annual production is lower (e.g. India, Chapter 9). Often, temperatures are lower in the main season, lessening the impact of a temperature rise. Nevertheless, in many of the sites (e.g. the Philippines, Chapter 13), spikelet sterility places a major limitation on higher yields in the main season, and ways of offsetting its effect would be advantageous.

Fortunately, considerable variation exists between varieties in tolerance to these high temperatures (Satake & Yoshida, 1978), and it was shown for the site at Madurai, India, that if the relationship of spikelet sterility to temperature was shifted 2°C higher, representing genotypic selection for this character, then the detrimental effect of the predicted scenarios could be offset completely, even for the 'worst-case' GISS scenario used in this example (Fig. 7.12). As this shift in response was well within the genotypic variation observed in current varieties, some hope is offered to breeders in providing a direction for future crop improvement programs. Evidence suggests that 'tolerant' varieties avoid the high day temperatures by flowering earlier in the morning (Imaki *et al.*, 1987).

Adjusting planting dates is another strategy which future farmers may use to adapt to a changed climate, particularly at high latitudes where a rise in temperature will lengthen the period in which rice can be grown. It was shown for a site in northern China that by advancing the planting date of rice by 30 days, which is not possible under the current climate, yield increases of 43% might be obtained (Fig. 7.14), due to both the fertilizing effect of an increased CO_2 concentration and to the panicle formation and grain-filling periods experiencing higher solar radiation. However, changing planting dates was not possible at all sites. At Madurai in southern India, for example, significant yield decreases under the GISS scenario were predicted if current planting dates were used, due to high spikelet sterility (Fig. 7.15). It was possible to offset these yield reductions if planting was delayed for about one month, but in this case, planting of the second crop in the sequence was also delayed, exposing it to higher temperatures and high spikelet sterility. Thus, current main season yields may be able to be maintained at such sites, but only at the expense of a drop in total annual productivity. Similarly, although not considered in this study, water availability may also place a constraint on the degree of flexibility of adjusting planting dates.

Using longer maturing varieties to take advantage of the increased growing seasons at higher latitudes, as suggested by some authors, was shown to not necessarily give higher yields. When hypothetical varieties with an increasingly longer basic vegetative period, and hence longer time to maturity, were 'grown' at Shenyang in northern China with the planting date advanced by one month, predicted yields actually declined as crop duration increased (Fig. 7.13). The reason for this was that extending the duration delayed the panicle formation and grain-filling periods until later in the season when incident solar radiation and temperatures are lower.

14.3   Limitations of the Analysis

The results of the present simulation study depend on the many assumptions built into the models used. In the case of the crop models, many of the effects of increased CO_2 levels have been derived from a limited number of experiments, often in controlled environments, and are, therefore, not fully tested, particularly for field conditions. In addition, other factors such as a current lack of knowledge of processes involved, and limited data to represent large geographical areas, all set limitations on the validity of the analysis. The following is a discussion of some of the major limitations of the analysis in this report.

In the agroclimatic evaluation of changes in the area and distribution of rice-growing environments (Chapter 6), no account is taken of the suitability of soils or the presence of geographical barriers, such as the Himalayan mountain ranges. Although zones based on changes in temperature and precipitation may move northwards, the associated rice ecosystems may not be able to. Similarly, the PPT–PET difference (see Chapter 6) is not necessarily the best index of water availability, particularly in irrigated ecosystems. Significant transfers of water along rivers and irrigation systems mean that adequate water may be available in areas with 'low' PPT–PET values, brought from areas with 'high' PPT–PET values but where rice is not grown for other reasons. The use of a linear temperature sum approach must also be interpreted with care; crop development has a distinct optimum (around 30°C, Fig. 3.2) beyond which there is no further shortening in crop duration. The relation is, therefore, not linear, particularly in the elevated temperature range considered in this study; in this case, predictions for areas with high temperatures may be in some error.

With the GCMs, the most significant limitations are their poor spatial resolution, inadequate coupling of atmospheric and oceanic processes, poor simulation of cloud processes (Mearns, 1990), and inadequate representation of the biosphere and its feedbacks (Dickinson, 1989). Similarly, most available runs are in equilibrium rather than transient, which is of critical importance for simulating mitigation options. Most GCMs have difficulty in even describing current climate adequately, particularly in relation to precipitation (Bachelet *et al.*, 1993), let alone climates several decades hence. It seems, therefore, that GCMs can at best be used to suggest the likely direction and rate of change of future climates. Current GCMs are also not able to predict changes in the variability of the weather or the frequency of catastrophic events such as hurricanes, floods, or even the intensity of monsoons, all of which can be just as, or more, important in determining crop yields as the average climatic data.

The crop models, also, are subject to many uncertainties. For example, most of the relationships relating the effect of temperature and CO_2 on plant processes are derived from experiments in which the crop's environment was changed for only part of the season; acclimation of the crop to changes in its environment is not taken account of in the model. Studies have shown that in some crops growing under enhanced CO_2 conditions, there is initially a large response, but over time, this response declines and approaches that of crops growing under current CO_2 levels (Peet, 1986). Higher CO_2 levels have also been found to hasten development rates (Baker *et al.*, 1990b), but, in current versions of both models, phenological development is taken to be independent of growth processes. The difference in the way both phenological development and crop respiration are described in the two models (Section 7.4.1) also pinpoints the need for further research to clarify the effect on crop processes of temperatures in excess of 30°C.

The assumption that changes in the potential yields represent the same proportional changes at all levels of management is another possible limitation to the validity of the results. Horie *et al.* (1992) have shown that simulation of potential production coupled with a multiplier to account for the level of technology used can reasonably accurately explain yearly variation in rice yields due to temperature and solar radiation levels, which would support this assumption. However, it has been suggested that when some resources are limited, the plants' response to higher CO_2 levels could be reduced (Cure & Acock, 1986), although in Jansen's (1990) study, greater proportional increases in yield were obtained under nitrogen-limited conditions. It is, therefore, not clear in which direction any possible error might occur, and until further information on this is forthcoming, the assumption appears reasonable. Potential production also assumes that water is not limiting; this is certainly true for irrigated rice, and to some extent, rainfed rice, which together represent 93% of the total world rice production. The effect of doubled-CO_2 levels on local water balances are uncertain; while the higher temperatures of a changed climate are likely to increase evaporation, all of the GCM scenarios also predict an increase in average global precipitation, so that variables such as soil moisture, runoff, and lake levels may not substantially change. Again, until more reliable predictions of the effect of climate change on water availability are obtained, the assumption that water is not limiting seems justified.

Another source of uncertainty in the results lies in the sparseness of weather data sites in some countries; large areas in both India and China, for example, are represented by only a few stations. While an attempt was made to stratify these areas into agroecological zones (AEZs), it is not known to what extent weather conditions are homogenous within a zone. Comparison of the changes in different countries, but the same AEZ (e.g. Thailand and Myanmar, both Zone 2, Tables 7.9 and 7.10), suggest that there could be significant variability in climate within a designated zone. However, until more high-quality weather data become available to enable a more detailed coverage, estimates based on the current data cannot be more accurate.

Nevertheless, despite these limitations, we feel that the current study marks significant progress in our understanding of how future climates are likely to affect rice production in the Asian region. We have built up an extensive database of daily weather data from a large number of sites in the Asian region, along with information on AEZs and current planting practices in the respective countries. This, combined with the use of two crop simulation models, in addition to three GCMs and various 'fixed-change' scenarios, provides a comprehensive evaluation of the sensitivity of the rice production system to changes in climatic parameters. We would like to reiterate the point made in the introductory chapter of this report, that the use of simulation models to predict the likely effects of climate change on crop production is an evolving science, and we hope that this study can be used as a baseline for future studies, in which some of the current limitations are addressed, so that increasingly better predictions can be made. Other levels of production, where the influences of water, nutrients, and pests and diseases are included in the models, are the next step that should be taken.

In this way, such studies may help to reduce the general reluctance by many governments to take action to mitigate both the rate of change of climate and the detrimental effects these changes may have, in order that we, as a species, may preserve an environment favorable to ourselves and the other species with whom we share this planet.

References

Achanta, A.N., 1993. An assessment of potential impact of global warming on Indian rice production. In: *The Climate Change Agenda: an Indian Perspective* (A.N. Achanta, ed.). Tata Energy Research Institute, New Delhi, India, pp. 113–141.

Acock, B. & L.H. Allen Jr, 1985. Crop responses to elevated carbon dioxide concentrations. In: *Direct Effects of Increasing Carbon Dioxide on Vegetation* (B.R. Strain & J.D. Cure, eds). United States Department of Energy DOE/ER-0238, pp. 53–98.

Akita, S., 1980. Studies on the differences in photosynthesis and photorespiration among crops. II. The differential responses of photosynthesis, photorespiration and dry matter production to carbon dioxide concentration among species. *Bulletin of the National Institute of Agricultural Sciences (Series D)* 31:59–94.

Amthor, J.S., 1984. The role of maintenance respiration in plant growth. *Plant, Cell and Environment* 7:561–569.

Angström, A., 1924. Solar and terrestrial radiation. *Quart. J. Roy. Met. Soc.* 50:121–126.

Anonymous, 1978. Report on the agro-ecological zones project, Vol. 3, Methodology and results for South and Central America. *Food and Agricultural Organization of the United Nations, World Soil Resources Report* 48, Rome.

Asuncion, M.T., 1971. An analysis of potential evapotranspiration and evaporation records at Los Baños, Laguna, The Philippines. University of the Philippines, Diliman, Quezon City.

Bachelet, D., D. Brown, M. Böhm & P. Russell, 1992. Climate change in Thailand and its potential impact on rice yield. *Climatic Change* 21:347–366.

Bachelet, D., J. Van Sickle & C.A. Gay, 1993. The impacts of climatic change on rice yield: evaluation of the efficacy of different modelling approaches. In: *Systems Approaches for Agricultural Development* (F.W.T. Penning de Vries, P.S. Teng & K. Metselaar, eds). Kluwer Academic Publishers, Dordrecht, The Netherlands, pp. 145–174.

Baker, J.T., L.H. Allen Jr & K.J. Boote, 1990a. Growth and yield responses of rice to carbon dioxide concentration. *J. Agric. Sci. (Camb.)* 115:313–320.

Baker, J.T., L.H. Allen Jr, K.J. Boote, P. Jones & J.W. Jones, 1990b. Developmental responses of rice to photoperiod and carbon dioxide concentration. *Agric. For. Meteorol.* 50:201–210.

Baker, J.T., L.H. Allen Jr, K.J. Boote, A.J. Rowland-Bamford, R.S. Waschmann, J.W. Jones, P.H. Jones & G. Bowes, 1990c. Temperature effects on rice at elevated CO_2 concentration. *1989 Progress Report of Response of Vegetation to Carbon Dioxide*, No. 060. Plant Stress and Protection Research Unit, USDA-ARS, University of Florida, Gainesville, 70 pp.

Baker, J.T., L.H. Allen Jr & K.J. Boote, 1992a. Temperature effects on rice at elevated CO_2 concentration. *J. Exp. Bot.* 43:959–964.

Baker, J.T., L.H. Allen Jr & K.J. Boote, 1992b. Response of rice to carbon dioxide and temperature. *Agric. For. Meteorol.* 60:153–166.

Barnett, K.H. & R.B. Pearce, 1983. Source-sink ratio alteration and its effect on physiological parameters in maize. *Crop Sci.* 23:294–299.

Bureau of Agricultural Statistics, 1990. *Rice Statistics Handbook 1989.* Department of Agriculture, Quezon City, The Philippines.

CES, 1992a. Annual Research Report of Crop Experiment Station. CES, Suweon, South Korea.

CES, 1992b. Annual Report of Regional Adaptivity Yield Trials. Moon Jung Dang, Suweon, South Korea.

Cess, R.D. & G.L. Potter, 1987. Exploratory studies of cloud radiative forcing with a general circulation model. *Tellus* 39(a):460–473.

CNRRI, 1988. *Regional Zonation of Rice Cropping in China.* China National Rice Research Institute, Zhejiang Publishing Press of Science and Technology, Hangzhou (Chinese with English summary).

Cohen, S.J., 1990. Bringing the global warming issue closer to home: the challenge of regional impact studies. *Bull. Amer. Meteorol. Soc.* 71:520–526.

Cramer, W.P. & A.M. Solomon, 1993. Climatic classification and future global redistribution of agricultural land. *Climate Res.* 3:97–110.

Crosson, P.R., 1989. Climate changes, American agriculture and natural resources: discussion. *Amer. J. Agric. Econ.* 71:1283–1285.

Cure, J.D. & B. Acock, 1986. Crop responses to carbon dioxide doubling: a literature survey. *Agric. For. Meteorol.* 38:127–145.

Decker, W.L., V.K. Jones & R. Achutuni, 1988. *The Impact of Climatic Change from Increased Atmospheric Carbon Dioxide on American Agriculture.* TR0-031, U.S. Dept. of Energy, Washington, DC.

Dickinson, R.E., 1989. Uncertainties of estimates of climatic change: a review. *Climatic Change* 15:5–13.

Diepen, C.A. van, H. van Keulen, F.W.T. Penning de Vries, I.G.A.M. Noy & J. Goudriaan, 1987. Simulated variability of wheat and rice yields in current weather conditions and in future weather when ambient CO_2 had doubled. *Simulation Report* CABO-TT 14, CABO, Wageningen, The Netherlands.

Ehleringer, J. & R.W. Pearcy, 1983. Variation in quantum yield for CO_2 uptake among C_3 and C_4 plants. *Plant Physiol.* 73:555–559.

Escaño, C.R. & L.V. Buendia, 1989. *Climate Impact Assessment for Agriculture in The Philippines.* Crops Research Division, The Philippine Council for Agriculture, Forestry, and Natural Resources Research and Development, Los Baños, Laguna, The Philippines.

FAO, 1987. *Agroclimatological data for Asia, FAO Plant Production and Protection Series* No. 25, Rome, Italy.

Fischer, G., K. Frohberg, M.L. Parry & C. Rosenzweig, 1994. Climate change and world food supply, demand and trade. *Global Environmental Change* 4:7–23.

Fischer, R.A., 1985. Number of kernels in wheat crops and the influence of solar radiation and temperature. *J. Agric. Sci. Camb.* 105:447–461.

Frère M. & G.F. Popov, 1979. Agrometeorological crop monitoring and forecasting. *Plant Production and Protection Paper* 17, FAO, Rome, 64 pp.

Gallagher, J.N. & P.V. Biscoe, 1978. Radiation absorption, growth and yield of cereals. *J. Agri. Sci.* 91: 47–60.

Gao L.Z., Z.Q. Jin, Y. Huang & L.Z. Zhang, 1992. Rice clock model – a computer model to simulate rice development. *Agric. For. Meteorol.* 60:1–16.

Gates, W.L., J.F.B. Mitchell, G.J. Boer, U. Cubash & V.P. Meleshko, 1992. Climate modeling, climate prediction and model validation. In: *Climate Change 1992. The Supplementary Report to the IPCC Scientific Assessment* (J.T. Houghton, B.A. Callander & S.K. Varney, eds). Cambridge University Press, Cambridge, pp. 99–134.

Gifford, R.M., 1979. Growth and yield of carbon dioxide-enriched wheat under water-limited conditions. *Aust. J. Plant Physiol.* 6:367–378.

Godwin, D., U. Singh, J. T. Ritchie & E.C. Alocilja, 1993. *A User's Guide to CERES-Rice.* International Fertilizer Development Center, Muscle Shoals, Alabama, 131 pp.

Goudriaan, J., 1977. *Crop Micrometeorology: A Simulation Study.* Pudoc, Wageningen, 257 pp.

Goudriaan, J., 1982. Some techniques in dynamic simulation. In: *Simulation of Plant Growth and Crop Production* (F.W.T. Penning de Vries & H.H. van Laar, eds). Pudoc, Wageningen, pp. 66–84.

Goudriaan, J., 1986. A simple and fast numerical method for the computation of daily totals of canopy photosynthesis. *Agric. For. Meteorol.* 43:251–255.

Goudriaan, J. & H.H. van Laar, 1978. Relations between leaf resistance, CO_2 concentration and CO_2 assimilation in maize, beans, lalang grass and sunflower. *Photosynthetica* 12:241–249.

Goudriaan, J. & H.H. van Laar, 1994. *Modeling Crop Growth Processes.* Kluwer Academic Publishers, Dordrecht, The Netherlands, 238 pp.

Government of India, 1989. *Agroclimatic Regional Planning: An Overview.* Planning Commission, New Delhi. Government Printing Press, New Delhi, India.

Grotch, S.L., 1988. *Regional Intercomparisons of General Circulation Model Predictions and Historical Climate Data.* US Dept of Energy Report TR041, Carbon Dioxide Research Division, Washington, DC DOE/NBB-0084, 291 pp.

Gutowski, W.J., D.S. Gutzler, D. Portmand & W.C. Wang, 1988. *Surface Energy Balance of Three General Circulation Models: Current Climate and Response to Increasing Atmospheric CO_2.* US Dept of Energy, Carbon Dioxide Research Division, Washington, DC DOE/ER/60422-H1, 119 pp.

Hammer, G.L., R.L. Vanderlip, G. Gibson, L.J. Wade, R. G. Henzell, D.R. Younger, J. Warren & A.B. Dale, 1989. Genotype-by-environment interaction in grain sorghum. II. Effects of temperature and photoperiod on ontogeny. *Crop Sci.* 29:376–384.

Hansen, J., I. Fung, A. Lacis, D. Rind, S. Lebedeff, R. Ruedy & G. Russell, 1988. Global climate changes as forecast by Goddard Institute for Space Studies three-dimensional model. *J. Geophys. Res.* 93:9341–9364.

Hatch, W., 1986. *Selective Guide to Climatic Sources. Key to Meteorological Records Documentation.* National Climatic Data Center, Asheville, North Carolina.

Higgins, G.M., A.H. Kassam, H.T. van Velthuizen & M.F. Purnell, 1987. Methods used by FAO to estimate environmental resources, potential outputs of crops, and population-supporting capacities in developing nations. In: *Agricultural Environments: Characterization, Classification and Mapping* (A. H. Bunting, ed.). CAB International, Wallingford, pp. 171–183.

Hodges, T., 1991. Crop growth simulation and the role of phenological models. In: *Predicting Crop Phenology* (T. Hodges, ed.). CRC Press, Boca Raton, Florida, pp. 3–13.

Horie, T., 1987. A model for evaluating climatic productivity and water balance of irrigated rice and its application to Southeast Asia. *Southeast Asian Studies* 25:62–74.

Horie, T., 1988. The effects of climatic variations on agriculture in Japan. 5: The effects on rice yields in Hokkaido. In: *The Impact of Climatic Variations on Agriculture, Vol. 1: Assessments in Cool Temperature and Cold Regions* (M.L. Parry, T.L. Carter & N.T. Konijn, eds). Kluwer Academic Publishers, Dordrecht, The Netherlands, pp. 809–826.

Horie, T., 1991. Model analysis of the effect of climatic variation on rice yield in Japan. In: *Climatic Variation and Change: Implications for Agriculture in the Pacific Rim* (S. Geng & C.W. Cady, eds). University of California, Davis, California.

Horie, T., 1993. Predicting the effects of climatic variation and effect of CO_2 on rice yield in Japan. *J. Agr. Meteor. (Tokyo)* 48:567–574.

Horie, T. & H. Nakagawa, 1990. Modeling and prediction of development process in rice I. Structure and method of parameter estimation of a model for simulating developmental process toward heading. *Jap. J. Crop Sci.* 59:687–695.

Horie, T. & T. Sakuratani, 1985. Studies on crop–weather relationship model in rice. I. Relation between absorbed solar radiation by the crop and the dry matter production. *J. Agr. Meteor. (Tokyo)* 40:331–342.

Horie, T., C.T. de Wit, J. Goudriaan & J. Bensink, 1979. A formal template for the development of cucumber in its vegetative stage. *Proc. KNAW, Ser. C* 82(4):433–479.

Horie, T., M. Yajima & H. Nakagawa, 1992. Yield forecasting. *Agric. Systems* 40:211–236.

Houghton, J.T., B.A. Callander & S.K. Varney, 1992. *Climatic Change 1992. The Supplementary Report to the IPCC Scientific Assessment.* Cambridge University Press, Cambridge, 200 pp.

Houghton, J.T., G.J. Jenkins & J.J. Ephraums (eds), 1990. *Climatic Change. The IPCC Scientific Assessment.* Cambridge University Press, Cambridge, 365 pp.

Huke, R.E., 1982a. *Agroclimatic and Dry-Season Maps of South, Southeast, and East Asia.* International Rice Research Institute, Los Baños, The Philippines.

Huke, R.E., 1982b. *Rice Area by Type of Culture: South, Southeast, and East Asia.* International Rice Research Institute, Los Baños, The Philippines, 32 pp.

Hulme, M., T. Wigley, T. Jiang, Z.-C. Zhao, F. Wang, Y. Ding, R. Leemans & A. Markham, 1992. *Climate Change due to the Greenhouse Effects and its Implications for China.* World Wide Fund for Nature, Gland, Switzerland, 58 pp.

Hunt, L.A., J.W. Jones, G. Hoogenboom, D.C. Godwin, U. Singh, N. Pickering, P.K. Thornton, K.J. Boote & J.T. Ritchie, 1994. *General Input and Output File Structures for Crop Simulation Models.* CoData, International Council of Scientific Unions, pp. 35–72.

Imai, K., D.F. Colman & T. Yanagisawa, 1985. Increase of atmospheric partial pressure of carbon dioxide and growth and yield of rice (*Oryza sativa* L.). *Jap. J. Crop Sci.* 54:413–418.

Imaki, T., S. Tokunaga & S. Obara, 1987. High temperature-induced spikelet sterility of rice in relation to flowering time. *Jap. J. Crop Sci.* 56:209–210 (in Japanese).

IRRI, 1991. *World Rice Statistics 1990.* International Rice Research Institute, Los Baños, The Philippines, 320 pp.

IRRI, 1993. *IRRI Rice Almanac 1993–1995.* International Rice Research Institute, Los Baños, The Philippines, 142 pp.

Islam, M.S. & J.I.L. Morison, 1992. Influence of solar radiation and temperature on irrigated rice grain yield in Bangladesh. *Field Crops Res.* 30:13–28.

Jansen, D. M., 1990. Potential rice yields in future weather conditions in different parts of Asia. *Neth. J. Agric. Sci.* 38:661–680.

Japanese Ministry of Agriculture, Forestry and Fisheries, various years. *Crop Statistics.* MAFF, Tokyo (in Japanese).

Jin Zhiqing, Ge Daokou, Chen Hua, Fang Juan & Zheng Xilian, 1993. Impacts of climate change on rice production and strategies for adaptation in southern China. In: *Proc. Int. Symp. on Climate Change, Natural Disasters and Agricultural Strategies* (Gao Liangzhi, Wu Lianhai, Zheng Dawei & Han Xiangling, eds). May 26–29, 1993. Beijing. Chinese Meteorological Press, Beijing.

Joyce, L.A., M.A. Fosberg & J.M. Comanor, 1990. Climatic change and America's Forests. *USDA Forest Service General Technical Report* RM-187. USDA, Washington, DC, 12 pp.

Karim, Z., M. Ahmed, S.G. Hussain & Kh.B. Rashid, 1991. *Impact of Climate Change on the Production of Modern Rice in Bangladesh.* Bangladesh Agricultural Research Council, Dhaka, Bangladesh.

Keeling, C.D., A.F. Carter & W.G. Mook, 1984. Seasonal, latitudinal, and secular variations in the abundance and isotopic ratios of atmospheric CO_2. *J. Geophys. Res.* 89:4615–4628.

Keulen, H. van & N.G. Seligman, 1987. *Simulation of Water Use, Nitrogen and Growth of a Spring Wheat Crop.* Pudoc, Wageningen, 310 pp.

Khanna, S.S., 1989. The agroclimatic approach. In: *Survey of Indian Agriculture.* The Hindu, Kasthuri Buildings, Madras 600 002, India, pp. 28–39.

Kim, H.Y., T. Horie, H. Nakagawa, K. Wada & T. Seo, 1992. Effects of elevated CO_2 concentration and high temperature on growth and nitrogen use efficiency of rice. In: *Proc. of the First Asian Crop Science Conference: Crop Production and Improvement Technology in Asia.* Korean Society of Crop Science, Seoul, South Korea, pp. 205–212.

Kimball, B.A., 1983. Carbon dioxide and agricultural yield. An assemblage and analysis of 430 prior observations. *Agron. J.* 75:779–788.

King, G.A., 1993. Conceptual approaches for incorporating climatic change into the development of forest management options for sequestering carbon. *Climate Res.* 3:61–78

Kiniry J.R., W.D. Rosenthal, B.S. Jackson & G. Hoogenboom, 1991. Predicting leaf development of crop plants. In: *Predicting Crop Phenology* (T. Hodges, ed.). CRC Press, Boca Raton, Florida, pp. 29–42.

Korean Meteorological Administration, 1991. *Climatological Standard Normals of Korea (1961–1990), Vol. 1: Daily and 10-day Normals.* Korean Meteorological Administration, Seoul, South Korea, 447 pp.

Korean Ministry of Agriculture, Forestry and Fisheries, 1992. *Statistics Book 1980–1992.* Korean Ministry of Agriculture, Forestry and Fisheries, Seoul, South Korea.

Kraalingen, D.W.G. van, W. Stol, P.W.J. Uithol & M.G.M. Verbeek, 1991. *User Manual of CABO/TPE Weather System.* CABO/TPE, Wageningen, The Netherlands.

Kropff, M.J. & H.H. van Laar, 1993. *Modelling Crop–Weed Interactions.* CAB International, Wallingford, UK, and International Rice Research Institute, Manila, The Philippines, 274 pp.

Kropff, M.J., H.H. van Laar, & H.F.M. ten Berge (eds), 1993. *ORYZA1: A Basic Model for Irrigated Lowland Rice Production.* Simulation and Systems Analysis for Rice Production (SARP) Publication, International Rice Research Institute, Los Baños, The Philippines.

Kropff, M.J., K.G. Cassman, F.W.T. Penning de Vries & H.H. van Laar, 1993. Increasing the yield plateau in rice and the role of global climate change. *J. Agr. Met.* 48:795–798.

Kumar, K.R, K.K. Kumar & G.B. Pant, 1994. Diurnal asymmetry of surface temperature trends over India. *Geophysical Research Letters* 21:677–680.

Kutzbach, J.E., 1987. Model simulations of the climatic patterns during the deglaciation of North America. In: *North America and Adjacent Oceans during the Last Deglaciation* (W.F. Ruddiman & H.E. Wright Jr, eds). Geology of North America, Geological Society of America, V. K-3, Boulder, Colorado.

Laar, H.H. van, J. Goudriaan & H. van Keulen (eds), 1992. *Simulation of Crop Growth for Potential and Water-limited Production Situations (as Applied to Spring Wheat).* CABO-TT Simulation Reports 27, Centre for Agrobiological Research, Wageningen, 72 pp.

Lansigan, F.P. & J.L Orno, 1991. Impact of climate change on rice yield variability in the Philippines. In: *Climate Variations and Change: Implications for Agriculture in the Pacific Rim* (S. Geng & C.W. Cady, eds). University of California, Davis, California, pp. 177–184.

Lansigan, F.P., 1993. Evaluating the effects of anticipated climate change on rice production in the Philippines. *J. Agr. Met.* 48:779–782.

Leemans, R. & W.P. Cramer, 1992. IIASA database for mean monthly values of temperature, precipitation, and cloudiness on a global terrestrial grid. In: *Global Ecosystems Database Version 1.0: Disc A, Documentation Manual* (J.J. Kineman & M.A. Ochrenschall, eds). US Dept of Comm./National Oceanic and Atmospheric Administration, National Geophysical Data Center, Boulder, Colorado.

Leemans, R. & A.M. Solomon, 1993. Modeling the potential change in yield and distribution of the earth's crops under a warmed climate. *Climate Res.* 3:79–96.

Lemon, E.R., 1983. *CO_2 and Plants: The Response of Plants to Rising Levels of Atmospheric Carbon Dioxide.* Westview Press, Boulder, Colorado.

Louwerse, W., L. Sibma & J. van Kleef, 1990. Crop photosynthesis, respiration and dry matter production of maize. *Neth. Journal Agric. Sci.* 38:95–108.

Major D.J. & J.R. Kiniry, 1991. Predicting daylength effects on phenological processes. In: *Predicting Crop Phenology* (T. Hodges, ed.). CRC Press, Boca Raton, Florida, pp. 15–28.

Manalo, E.B., 1977a. *Agroclimatic Survey of Bangladesh.* Bangladesh Rice Research Institute, International Rice Research Institute, Los Baños, the Philippines. 361 pp.

Manalo, E.B., 1977b. *Agroclimatic Map of the Philippines.* International Rice Research Institute, Los Baños, The Philippines.

Marks, D.G., 1990. A continental-scale simulation of potential evapotranspiration for historical and projected doubled CO_2 climate conditions. In: *Biospheric Feedbacks to climate change, the sensitivity of regional trace gas emissions, evapotranspiration, and energy balance to vegetation redistribution* (H. Gucinski, D.G. Marks & D.P. Turner, eds). US EPA Technical Report. EPA/600/3-90/078, 273 pp.

Matsui, T. & T. Horie, 1992. Effects of elevated CO_2 and high temperature on growth and yield of rice. 2. Sensitive period and pollen germination rate in high temperature sterility of rice spikelets at flowering. *Jap. J. Crop Sci.* 61:148–149 (extra issue 1).

Matthews, R.B. & L.A. Hunt, 1993. *The GUMCAS v1.0 Cassava Growth Model Users' Manual.* Crop Science Simulation Series No. 2. University of Guelph, Canada, 129 pp.

Matthews, R.B. & L.A. Hunt, 1994. A model describing the growth of cassava (*Manihot esculenta* L. Crantz). *Field Crops Res.* 36:69–84.

Matthews, R.B., M.J. Kropff, T. Horie & D. Bachelet, 1995a. Simulating the impact of climate change on rice production in Asia. I. Crop model description and evaluation. *Agricultural Systems.*

Matthews, R.B., M.J. Kropff, T. Horie & D. Bachelet, 1995b. Simulating the impact of climate change on rice production in Asia. II. Predicted yield changes under various scenarios. *Agricultural Systems.*

Matthews, R.B., M.J. Kropff, T. Horie & D. Bachelet, 1995c. Simulating the impact of climate change on rice production in Asia. III. Adaptation options and conclusions of the study. *Agricultural Systems.*

Mearns, L.O., 1990. Future directions in climate modeling: a climate impacts perspective. In: *Climatic Change: Implications for Water and Ecological Resources* (G. Wall & M. Sanderson, eds). Proceedings of an International Symposium, Dept of Geography, University of Waterloo, Canada, pp. 51–58.

Mearns, L.O., 1991. Climate variability: possible changes with climatic change and impacts on crop yields. In: *Symposium on the Impacts of Climatic Change and Variability on the Great Plains* (G. Wall, ed.). Dept of Geography, University of Waterloo, Canada, pp. 147–157.

Min Shaokai & Fei Huailin, 1984. Yield constraints and production prospects of rice in China. In: *International Crop Science Symposium on Potential Production and Yield Constraints of Rice in East Asia*, October 17–20, 1984, Fukuoka, Japan. Crop Science Society of Japan.

Mohamed, H.A., J.A. Clark & C.K. Ong, 1988. Genotypic differences in the temperature responses of tropical crops. I. Germination characteristics of groundnut (*Arachis hypogaea* L.) and pearl millet (*Pennisetum typhoides* S. & H.). *J. Exp. Bot.* 39:1121–1128.

Monsi, M. & T. Saeki, 1953. Uber den Lichtfaktor in den Pflanzengesellschaften und seine bedeutung fur die Stopffproduktion. *Jap. J. Bot.* 15:22–52.

Monteith, J.L., 1969. Light interception and radiative exchange in crop stands. In: *Physiological Aspects of Crop Yield* (J.D. Eastin, F.A. Haskins, C.Y. Sullivan & C.H.M. van Bavel, eds). American Society of Agronomy, Crop Science Soc. of America, Madison, Wisconsin, pp. 89–111.

Monteith, J.L., 1977. Climate and efficiency of crop production in Britain. *Philosoph. Trans. R. Soc. Lon.* B 281, pp. 277–294.

Munakata, K., 1976. Effects of temperature and light on the reproductive growth and ripening of rice. In: *Proc. of the Symp. on Climate and Rice.* International Rice Research Institute, Los Baños, The Philippines, pp. 187–210.

Nakagawa, H., T. Horie, J. Nakano, H.Y. Kim, K. Wada & M. Kobayashi, 1993. Effect of elevated CO_2 concentration and high temperature on growth and development of rice. *J. Agric. Meteor.* 48:799–802.

NIAS, 1975. *Climatic Change and Crop Production.* Misc. Report, Division of Meteorology, National Institute of Sciences, Tokyo, Japan, 40 pp. (in Japanese).

Nicholls, N., 1990. Predicting the El Niño-Southern Oscillation. *Search* 21:165–167.

Nonhebel, S., 1993. The importance of weather data in crop growth simulation models and assessment of climatic change effects. PhD thesis, Agricultural University Wageningen, The Netherlands, 144 pp.

Okada, M., 1991. Variations of climate and rice production in Northern Japan. In: *Climate Variations and Change: Implications for Agriculture in the Pacific Rim* (S. Geng & C. W. Cady, eds). University of California, Davis, California.

Oldeman, L.R., 1975. An agroclimatic map of Java and Madura. *Contr. Cent. Res. Inst. Agric., Bogor.* No. 17, 22 pp.

Oldeman, L.R., 1977. Climate of Indonesia. In: *Proceedings of the 6th Asian–Pacific Weed Science Society Conference, Jakarta, Indonesia.* Vol. 1, pp. 14–30. Asian–Pacific Weed Science Society.

Oldeman, L.R., I. Las & S. N. Darwis, 1979. An agroclimatic map of Sumatra. *Contr. Centr. Res. Inst. Agric., Bogor.* No. 52, 35 pp.

Oldeman, L.R., D.V. Seshu & F.B. Cady, 1986. Response of rice to weather variables. Trials conducted by cooperating scientists from 16 countries. Report of an IRRI/WMO Special Project, International Rice Research Institute, Los Baños, The Philippines.

Oldeman, L.R., D.V. Seshu & F.B. Cady, 1987. Response of rice to weather variables. In: *Weather and Rice,* IRRI, Los Baños, The Philippines, pp. 5–39.

Ozawa, Y., 1962. *Climatic Zoning of Japan for Land Use.* Chiri 7:300–364 and 432–438. Kokinshoin, Tokyo (in Japanese).

Panturat, S. & A. Eddy, 1990. Some impacts of rice yield from changes in the variance of precipitation. *Clim. Bull.* 24:16–26.

Parry, M.L. & T.R. Carter, 1988. The assessment of effects of climatic variations on agriculture: aims, methods and summary of results. In: *The Impact of Climatic Variations on Agriculture* (M.L. Parry, T.R. Carter & N.T. Konijn, eds). Kluwer Academic Publishers, Dordrecht, The Netherlands, pp. 11–95.

Parry, M.L., T.R. Carter & N.T. Konijn (eds), 1988. *The Impact of Climatic Variations on Agriculture.* Kluwer Academic Publishers, Dordrecht, The Netherlands.

Peet, M.M., 1986. Acclimation to high CO_2 in monoecious cucumbers. I. Vegetative and reproductive growth. *Plant Physiology* 80:59–62.

Penman, H., 1948. Natural evaporation from open water, bare soil and grass. *Proc. Royal Soc. Lond. Ser. A* 193:120–146.

Penning de Vries, F.W.T., 1975. The costs of maintenance processes in plant cells. *Annals of Botany* 39:77–92.

Penning de Vries, F.W.T., 1993. Rice production and climate change. In: *Systems Approaches for Agricultural Development* (F.W.T. Penning de Vries, P.S. Teng & K. Metselaar, eds). Kluwer Academic Publishers, Dordrecht, The Netherlands, pp. 175–192.

Penning de Vries, F.W.T. & H.H. van Laar (eds), 1982. *Simulation of Plant Growth and Crop Production.* Pudoc, Wageningen, 308 pp.

Penning de Vries, F.W.T., A.H.M. Brunsting & H.H. van Laar, 1974. Products, requirements and efficiency of biosynthesis: a quantitative approach. *J. Theor. Biol.* 45:339–377.

Penning de Vries, F.W. T., D.M. Jansen, H.F.M. ten Berge & A. Bakema, 1989. *Simulation of Ecophysiological Processes of Growth in Several Annual Crops.* Pudoc, Wageningen and IRRI, Los Baños, 271 pp.

Penning de Vries, F.W.T., H. van Keulen & J.C. Alagos, 1990. Nitrogen redistribution and potential production in rice. In: *Proc. of the Int. Congress of Plant Physiology, 15–20 Feb. 1988, New Delhi, India* (S.K. Sinha, P.V. Sane, S.C. Bhargava & P.K. Aggarwal, eds), Volume 1. InPrint Exclusives, S-402, Greater Kailash-II, New Delhi, pp. 513–520.

Rao, K.N. & J.C. Das, 1971. *Weather and crop yields – rice* (Survey). IMD prepublished Scientific Report No. 137. Indian Meteorological Department, New Delhi.

RDA, 1981. *Special Reports of Chilling Injury of Rice in 1980, Korea. The Analysis of Chilling Injury and Possible Cultivation Methods.* Chungang Chulpansa, Suweon, South Korea.

Research Group of Evapotranspiration, 1967. Radiation balance of paddy field. *J. Agr. Meteor. (Tokyo)* 22:97–102.

Robinson, P.J. & P.L. Finkelstein, 1989. Strategies for the Development of Climate Scenarios. AREAL, US Environmental Protection Agency, Research Triangle Park, North Carolina, 73 pp.

Rosenberg, N.J., 1991. Process for identifying regional influences of and responses to increasing atmospheric CO_2 and climate change – the MINK project. *Report I – Background and Baselines*, TR052B, U.S. Dept of Energy, Washington, DC.

Rosenberg, N.J. & P.R. Crosson, 1991. The MINK project: a new methodology for identifying regional influences of, and responses to, increasing atmospheric CO_2 and climatic change. *Environ. Conserv.* 18:313–322.

Rosenzweig, C., M.L. Parry, G. Fischer & K. Frohberg, 1993. *Climate Change and World Food Supply.* Research Report No. 3, Environmental Change Unit, University of Oxford, UK.

Rosenzweig, C. & M.L. Parry, 1994. Potential impact of climate change on world food supply. *Nature* 367:133–138.

Satake, T. & S. Yoshida, 1978. High temperature-induced sterility in *indica* rice at flowering. *Jap. J. Crop Sci.* 47:6–17.

Schneider, S.H., P.H. Gleick & L.O. Mearns, 1989. Prospects for climatic change. In: *Climatic Change and U.S. Water Resources* (P. E. Waggoner, ed.). Wiley and Sons, New York.

Shibata. M., K. Sasaki and Y. Shimazaki, 1990. Effects of air-temperature and water-temperature at each stage of the growth of lowland rice. I. Effect of air-temperature and water-temperature on the percentage of sterile grains. *Proc. Crop Sci. Soc., Japan* 39, 401–408.

Shibles, R.M. & C.R. Weber, 1966. Interaction of solar radiation and dry matter production by various soybean planting patterns. *Crop Sci.* 6:55–59.

Sionit, N., H. Hellmers & B.R. Strain, 1980. Growth and yield of wheat under CO_2-enrichment and water stress. *Crop Sci.* 20:456–458.

Smith, J.B. & D.A. Tirpak (eds), 1989. *The Potential Effects of Global Climate Change on the United States, Vol. 1, Report to Congress.* EPA-230-05-89-050, U.S. Environmental Protection Agency, Washington, DC, 413 pp.

Solomon, A.M. & R. Leemans, 1990. Climatic change and landscape ecological response: Issues and analysis. In: *Landscape Ecological Impact of Climatic Change* (M.M. Boer, & R.S. de Groot, eds). IOS Press, Amsterdam, pp. 293–316.

Spitters, C.J.T., H.A.J.M. Toussaint & J. Goudriaan, 1986. Separating the diffuse and direct component of global radiation and its implications for modeling canopy photosynthesis. I. Components of incoming radiation. *Agric. For. Meteorol.* 38:217–229.

Spitters, C.J.T., H. van Keulen & D.W.G. van Kraalingen, 1989. A simple and universal crop growth simulator: SUCROS87. In: *Simulation and Systems Management in Crop Protection* (R. Rabbinge *et al.*, eds). Pudoc, Wageningen, pp.147–181.

Sreenivasan, P.S., 1980. Meteorological aspects of rice production in India. In: *Agrometeorology of Rice.* International Rice Research Institute, Los Baños, The Philippines, pp. 19–31.

Stern, R., J. Knock & H. Hack, 1990. INSTAT climatic guide. University of Reading, UK.

Strahler, A.N., 1975. *Physical Geography.* John Wiley and Sons, New York, 643 pp.

Supit, I., 1986. *Manual for Generation of Daily Weather Data.* Simulation Report CABO-TT No. 7. Centre for Agrobiological Research, Wageningen, The Netherlands, 46 pp.

Uchijima, T., 1976. Some aspects of the relation between low air temperature and sterile spikelets in rice plants. *J. Agric. Meteorol. (Tokyo)* 31:199–202.

UNEP Greenhouse Gas Abatement Costing Studies, 1992. *Analysis of Abatement Costing Issues and Preparation of a Methodology to Undertake National Greenhouse Abatement Costing Studies: Phase One Report.* UNEP Collaborating Centre on Energy and Environment, Risø National Laboratory, Denmark.

U.S. Army Corps of Engineers, 1988. *GRASS Users and Programmers Manual.* U.S. Army Corps of Engineers Construction Engineering Research Laboratory, Champaign, Illinois.

Vamadevan, V.K. & K.S. Murty, 1976. Influence of meteorological elements on productivity. In: *Rice Production Manual.* Indian Council of Agricultural Research, New Delhi, India, pp. 56–63.

Vergara, B.S. & T.T. Chang, 1985. The flowering response of the rice plant to photoperiod. A review of the literature. International Rice Research Institute, Los Baños, The Philippines, 61 pp.

Wang Futang, Wang Shili, Li Yuxiang & Zhong Meina, 1991. A preliminary modeling of the effects of climatic changes on food production in China. In: *Climate Variations and Change: Implications for Agriculture in the Pacific Rim* (S. Geng & C.W. Cady, eds). University of California, Davis, California, pp. 115–126.

Wetherald, R.T. & S. Manabe, 1988. Cloud feedback processes in a general circulation model. *J. Atmos. Sci.* 45:1397–1415.

Wilson, C.A. & J.F.B. Mitchell, 1987. A doubled CO_2 climate sensitivity experiment with a GCM including a simple ocean. *J. Geophys. Res.* 92:13315–13343.

Wit, C.T. de, R. Brouwer & F.W.T. Penning de Vries, 1970. The simulation of photosynthetic systems. In: *Proceeding of the IBP/PP, Technical Meeting, Trebon (1969).* Pudoc, Wageningen, pp. 47–60.

Wit, C.T. de, J. Goudriaan, H. H. van Laar, F.W.T. Penning de Vries, R. Rabbinge, H. van Keulen, L. Sibma & C. de Jonge, 1978. *Simulation of Assimilation, Respiration and Transpiration of Crops.* Pudoc, Wageningen, 141 pp.

WMO, 1989. *CLICOM Project (Climate Data Management System)*, World Climate Data Programme No. 299, Geneva. World Meteorological Organisation.

Wopereis, M., M. Kropff, J. Bouma, A. van Wijk & T. Woodhead (eds), 1994. *Soil Physical Properties: Measurement and Use in Rice-based Cropping Systems.* International Rice Research Institute, Los Baños, The Philippines, 111 pp.

Xiong Zhenmin, Cai Hongfa, Min Shaokai & Li Baochu, 1992. *Rice in China*. Chinese Publishing Press of Agricultural Science and Technology, Beijing (Chinese with English summary).

Yoshida, S., 1981. *Fundamentals of Rice Crop Science*. International Rice Research Institute, Los Baños, The Philippines, 269 pp.

Yoshida, S., 1983. Rice. In: *Potential Productivity of Field Crops under Different Environments*. International Rice Research Institute, Los Baños, The Philippines. pp.103–127.

Yoshida, S. & F.T. Parao, 1976. Climatic influence on yield and yield components of lowland rice in the tropics. In: *Climate and Rice*. International Rice Research Institute, Los Baños, The Philippines pp. 471–494.

Yoshino, M.M., T. Horie, H. Seino, H. Tsujii, T. Uchijima & Z. Uchijima, 1988a. The effects of climate variations on agriculture in Japan. In: *The Impact of Climate Variations on Agriculture. Vol. 1: Assessments in Cool Temperate and Cold Regions*. (M.L. Parry, T.R. Carter & N.T. Konijn, eds). Kluwer Academic Publishers, Dordrecht, The Netherlands, pp. 725–865.

Yoshino, M., Z. Uchijima & H. Tsujii, 1988b. The implications for agricultural policies and planning. In: *The Impact of Climatic Variations on Agriculture. Vol. 1: Assessments in Cool Temperate and Cold Regions*. (M.L. Parry, T.R. Carter & N.T. Konijn, eds). Kluwer Academic Publishers, Dordrecht, The Netherlands, pp. 853–868.

Zhou Yixing, 1991. Potential influences of the greenhouse effect on grain production in China. In: *Climate Variations and Change: Implications for Agriculture in the Pacific Rim* (S. Geng & C.W. Cady, eds). University of California, Davis, California, pp. 147–158.

Index

Note: page numbers in *italics* refer to figures and tables